ASSESSING AND ADDRESSING RISKS IN PACIFIC CONSTRUCTION SUPPLY CHAINS

NOVEMBER 2024

ASIAN DEVELOPMENT BANK

Contents

Abbreviations

ADB	Asian Development Bank
Austrade	Australian Trade and Investment Commission
COVID-19	coronavirus disease
DFAT	Australian Department of Foreign Affairs and Trade
MDB	multilateral development bank
MFAT	New Zealand Ministry of Foreign Affairs and Trade
PNG	Papua New Guinea
PRC	People's Republic of China
PRIF	Pacific Region Infrastructure Facility
UNESCAP	United Nations Economic and Social Commission for Asia and the Pacific
US	United States
WHS	workplace health and safety

Executive Summary

Construction projects in the Pacific are complex undertakings fraught with supply chain risks that are exacerbated by the region's extreme remoteness, vulnerability to disasters triggered by natural hazards, and small markets. In a vicious circle, weak domestic construction sectors, limited import and export demand, and insufficient infrastructure investment across the region leads to a shortage of the physical infrastructure necessary to support the development of robust supply chains capable of increasing transport efficiency and reducing associated costs. Instead, port facilities in the Pacific are often insufficient with shipping services infrequent and unreliable, leaving international contractors and suppliers with a heightened concern that infrastructure projects will be difficult to execute and not worth the risks inherent in their delivery.

This report examines the nature and sources of these risks in 14 Pacific island countries and shows that practical solutions exist, and experienced contractors have developed solutions to overcome the inherent logistic challenges to working successfully in the region. Focusing on the construction sector, the study is based on a review of relevant literature, an analysis of construction material import data, and extensive consultation involving interviews with 59 individuals from 52 private companies and government organizations active in construction supply chain ecosystems.

The study reveals that Pacific island countries face some common challenges, but they also differ considerably, even among main and outer islands in the same country. This reflects differing culture, governance, access to essential services, and infrastructure that shapes the achievement of successful outcomes on infrastructure projects. The following summarizes the key findings and recommendations for actions that can help overcome them.

Findings

Shipping Issues

Disruptions during the pandemic. Lockdowns during the coronavirus disease (COVID-19) pandemic upended shipping. Just 20% of ships arrived on time, ships skipped ports where cargo was insufficient, and costs of shipping and associated logistics services reached extremely high levels. This created significant challenges for

Pacific island countries, exposing their vulnerability to global shocks, with widespread local impacts. By December 2023, shipping prices normalized to below pre-COVID-19 levels. However, connectivity problems persist, especially in locations that lack adequate port infrastructure.

Vessel size limitations. Large cargo vessels are often optimized for specific cargo, typically containerized or bulk, creating challenges for construction projects that have precise shipping needs, particularly in relation to break-bulk cargo where damage during transit from rough seas is also a risk. Planning for transport in construction supply chains is thus challenging, complicating bidding processes and project timelines.

Port and infrastructure issues. Some Pacific island countries face substantial delays in offloading and clearing cargo due to inadequate port infrastructure, congestion, and the focus on nonconstruction commodities. As the demand for cargo increased post COVID-19, it revealed a lack of available shipping capacity. However, where additional shipping capacity has been added, this has often exacerbated already existing delays due to poor port infrastructure.

Material supply. Contractors often need to import a substantial proportion of the construction materials needed to deliver a project. Long lead times for delivery are exacerbated by frequently changing tariffs and taxes on goods entering some countries adding to financial risks and reducing contractor cash flow.

Labor availability. Quality management through use of qualified and experienced staff is crucial for project success however, skilled local labor is in short supply as they often migrate to other countries in the region for longer term, better-paying, jobs, exacerbating the labor shortage. In their absence, the use of imported labor is costly, and they are often reluctant to go to Pacific island countries for long durations due to their isolation, limited available services and, in some cases, volatile security environments. This situation is further compounded in some countries by immigration regulations.

Foreign exchange rate volatility. Rapid devaluation of local currencies often complicates business operations and increases supply chain risks. Difficulty in obtaining foreign currency and its related volatility can undermine revenues, raise costs, and reduce project profitability.

Insurance. Infrastructure projects normally require types and levels of insurance that local contractors struggle to obtain. In some cases, they need to rely on self-insurance or government support for small projects to bid, limiting competition. Large contractors are less impacted, with access to international insurance markets and an ability to insure at portfolio, rather than project, level. This topic is subject to another recent guide to insurable infrastructure in the Pacific.

Local governance. Local regulations and tax regimes, including their implementation, vary across the Pacific and poses significant risks. Contractors may face delays in regulatory approvals and customs clearances with some potentially using unofficial channels to try to accelerate the process. This increases both implementation and governance risks, and leads to perceptions of an uneven playing field.

A perceived focus on price and limited consequence of poor performance. Contractors view the current contracting models used in the Pacific as overly cost-focused. This is despite increased use of quality-based evaluation methods by multilateral development banks, meaning more communication is required. There are also calls for a more collaborative approach to risk-sharing and improved performance evaluation post-construction with consequences for poor quality delivery.

Short, Medium, and Long-term Supply Chain Outlook

In the short term (1–2 years), construction project supply chains in the Pacific hold promise. However, shipping capacity lost during COVID-19 has been slow to come back, which may add to constraints.

The medium term (3–10 years) outlook is less clear, with concerns about geopolitical tensions, insurance availability, and labor market insufficiencies (while larger islands have better prospects for upskilling, smaller islands will struggle to develop and retain skilled labor).

Long term (10+ years) risks include climate change, geopolitical uncertainty, labor migration, and supply scarcity. Contractors are concerned about their ability to source supplies and extreme weather events.

Recommendations

The observed supply chain risks can be overcome with proper mitigation measures. Indeed, the report repeatedly underscores the confidence among experienced contractors that these risks are manageable, though new entrants face higher vulnerability.

Planning, institutional capacity, and adaptability are essential. Participants must anticipate project needs, hedge foreign exchange rates and fuel costs, and utilize reliable shipping methods to transport materials to remote islands. Barges are often used for transportation due to their reliability but are expensive.

Vertical integration is another strategy employed by contractors to mitigate supply chain risks. By manufacturing building products locally and engaging local expertise, contractors reduce reliance on external subcontractors and support local employment. However, this approach can limit market entry for newcomers and reduce competition. It should also not be confused with another form of vertical integration whereby all construction materials and human resources are imported, hindering local economic development. Therefore, the modality of vertical integration adopted is important. Government enforcement of quality standards is crucial to ensure project longevity and sustainability.

Availability of insurance and incorporation of labor training activities are also important risk mitigation strategies as recently investigated by the Asian Development Bank as part of a recent related insurance study. Freight insurance helps protect against losses during transit, while labor training improves the skills and experience of local and international staff. Management systems and digitization of the supply chain improve forecasting, planning, and operations management.

Foreign exchange management through fixing procurement costs and using financial hedging can help contractors to mitigate foreign exchange risks. Standardizing price adjustment matrixes in contracts and continuing to allow contractors to be paid in multiple currencies can also assist.

Governments in the Pacific should employ various strategies to mitigate shipping risks, such as "exclusivity contracts" with shipping lines that require them to stop at their island and coordinated approaches to cargo handling among several islands. Adopting a regional focus to sourcing and logistics could mitigate some of the challenges faced. They should also enact any changes necessary to ensure that local legislation encourages contractors to do business in the country as robust governance and efficient government approvals are needed for all projects.

Clearly, supply chain risks arising from remote locations, weak local infrastructure, climate change, pandemics, and so on, make successful infrastructure projects a daunting challenge. However, contractors with experience in the region can manage supply chain risks effectively, making these projects financially and environmentally rewarding. New entrants often fail without local know-how, meaning that collaboration with experienced contractors is recommended for new market entrants to navigate the challenges successfully.

By incorporating the recommended short-term actions and fostering long-term partnerships, detailed in this report, governments, multilateral development banks, contractors, and local communities can build resilience, enhance project efficiency, and promote socioeconomic growth in the Pacific.

1 Introduction

Global and local supply chains help businesses, communities, and entire economies connect and function, especially in regional and remote regions such as the Pacific. Indeed, the coronavirus disease (COVID-19) pandemic and its aftermath underscored just how important functional supply chains are for every country and sector. This study examines the nature and source of risks associated with supply chains in the Pacific island countries, focusing on the construction sector.[1]

The analysis identifies significant and continuing vulnerabilities and possible approaches to manage the associated risks. All parties involved in a construction project must assess these risks early and identify suitable mitigation activities. The study also focuses on the frequent threats affecting the construction industry, including unforeseen delays and price increases related to supply chain problems and material shortages. The study reviews the major goods and infrastructure supply chains related to the construction sector across 14 Pacific island countries, focusing on how the identified risks may affect contracts funded by both multilateral development agencies and the countries themselves.

One associated issue, identified in project meetings with the Asian Development Bank (ADB) and the World Bank is that there is insufficient market competition and capacity to deliver infrastructure project pipelines, with contractors (construction companies) wary of problematic supply chain risks in the Pacific.

2 What Does Literature Tell Us About Supply Chain Risks?

Supply Chain Risks

In General

Construction supply chain risk in the Pacific is too "niche" to attract interest from peer reviewed academic journals, as revealed in the academic and "gray" literature.[2] However, academic literature does look, more generally, at supply chain risks (e.g., Rao and Goldsby 2009), their assessment (e.g., Choudhary et al. 2023), and strategies to overcome and mitigate them (e.g., Gurbuz et al. 2023). All are relevant to this study and explored in this report. It is generally accepted that supply chains consist of multiple actors and that supply chain design can help cost-effectively mitigate risks created by major disruptions, such as during the COVID-19 pandemic (e.g., Rinaldi and Bottani 2023; Zighan 2022). Typical strategies for risk management identified in this literature stream include the diversification of transport modes or an increase in stock and inventory, neither of which may be an option in the Pacific construction context due to limited warehousing capacity both in scale and scope. As technology advances, one supply chain risk often cited is the lack of labor force and skills (e.g., Merkert et al. 2023), a very relevant trend to Pacific countries. A further supply chain risk not specific to construction but often associated with the countries in the region is adaptation to climate change, particularly in maritime supply chains (e.g., Dyer 2017).

In Construction Projects

The academic literature on supply chain risks related to construction projects is broad, comprehensively classifying supply chain risk types and/or analyzing supply chain management processes in the construction sector. According to Pham et al. (2023), much of the literature focuses on identifying and assessing risks, while mitigating and monitoring risks has not yet received much attention. Project delays are one aspect given considerable attention (e.g., Panova and Hilletofth 2018). More recent academic literature on construction supply risks has focused on COVID-19 disruptions, but is limited to Ghana (Al-Mhdawi et al. 2023), Sri Lanka (Niroshana 2022), Iran, and the United States (US) (Rokooei, Alvanchi, and Rahimi 2022). This stream of literature has found that contractual risks have the strongest total effects on project success, followed by organizational risks (i.e., government or construction firms), financial market risks, construction workforce risks, and supply chain operation risks. This implies that supply chain risks are perceived as substantially less problematic than contractual issues and it will be interesting to see if this also applies in the Pacific countries.

---

2   Gray literature is information that is published informally or unpublished outside of traditional publishing and distribution channels. It is produced by governments, academia, industry, business, etc.

Interestingly, much of the gray literature, as discussed below, has focused on supply chain operations management (not including transportation), which is potentially the least important type of supply chain risk in construction projects. This report thus not only undertakes a more systematic review of the academic literature but also identifies supply chain risks in construction and develops a theoretical framework for mitigating the risks. We then apply this to the Pacific countries.

In addition to academic journals, considerable gray literature and consultancy reports discuss construction specific supply chain risks, including a very recent guide on supply chain management and procurement published by the World Bank (2023a). A multitude of risks have been identified, usually related to performance risks (e.g., skill shortages, financial stress, up- and downstream blockages), compliance risks (traditional obligations, emerging targets, regional and international differences) and uncertain world risks (COVID-19 pandemic, force majeure, criminality). Construction supply chains are not limited to materials but typically involve labor, plant, materials, and equipment as well as subcontractors and consultants, all of which are supported by complementary industries, such as freight and logistics (Thompson 2022). Other contributors point to the multifaceted complexity and temporary nature of construction supply chains, including location, politics, governance, and geography.

Supply Chain Risks in the Pacific Islands

Relatively comprehensive grey literature is available on supply chain risks in the Pacific, often commissioned by multilateral development banks (MDBs). Jephcott (2022) investigated supply chain disruption due to the COVID-19 pandemic and its economic impacts on business across the Pacific, identifying much reduced shipping line activity, disruption, and higher transport cost as the main impacts on Pacific business. Appendix 1 illustrates Pacific shipping routes and the interdependency of its countries on each other and on Australia and New Zealand. This is important for construction projects in the Pacific countries, as it shows that end-of-the-line countries will be impacted by ripple effects and that due to the circular nature of these routes, delays in one port may affect the arrival of the vessels in subsequent ports.

The United Nations Economic and Social Commission for Asia and the Pacific (UNESCAP 2022) investigated the sustainability and resilience of ports and maritime connectivity in the Pacific and identified issues relevant to supply chains. This includes, for example, a lack of regional network integration and coordination, leading to poor network efficiency, insufficient supply chain resilience, and high freight costs. UNESCAP highlighted aging and inadequate port infrastructure, the "build–neglect–rebuild" paradigm, vulnerability to climate change and natural hazards, difficulties monitoring and fairly comparing port productivity, poor governance and management at ports, and a need for a managed approach to investment in digitalization. Perhaps most relevant to construction, the UNESCAP report highlighted overly optimistic planning assumptions about economic viability, dependence on aid, limits on borrowing due to high risk of debt distress, undercharging for port services, and inadequate budgeting for maintenance and repairs. Sector governance and human resources and capacity were also mentioned, including a small pool of people, lack of access to training, low management capacity, and low participation of women in the sector.

The World Bank (2023b) detailed Pacific supply chain issues and strategies to overcome the challenges. The study was conducted in the aftermath of COVID-19, and thus related issues were still discussed. However, the focus had moved to risks related to ongoing economic and geopolitical uncertainties, compounded by long-standing supply chain difficulties in these logistically complicated and remote markets. The study highlighted volatility in fuel prices and commodity markets for raw material (e.g., steel) and high mobilization costs, as equipment, materials, and even labor are often transported or imported from abroad for many projects. It focused less on climate change, but it did mention risks related to extreme weather events and flooding, both of which can worsen as climate changes, which leaves many projects difficult to insure. While it sees global supply chain issues and localized risks as problematic, it also acknowledges that small project values contribute to the lack of bidders for donor funded projects in the Pacific countries and the associated high process observed to deliver these projects.

Very relevant to this study, the World Bank (2023b) identified a reduction in market interest in bids, as construction firms focus on more attractive projects that carry less perceived risk in the large pipeline of domestic infrastructure projects in their home markets. Importantly, the World Bank study focused on transport and logistics projects as this comprised approximately 83% of their project portfolio, by value, between 2018 and 2022 (Table 2). These figures are relatively consistent with ADB's lending program over the same period (67% in the transport and logistics sector, by value) meaning that supply chain risks and mitigation strategies identified in that report are very relevant to our study.

With an average ADB-funded project size of $21 million over 2018–2022, insurance availability may also be a project and supply chain risk, as shown by Infrastructure Risk Management and Insurance in the Pacific (2023), an ADB Pacific Region Infrastructure Facility (PRIF) report (PRIF 2024). The report found that for projects in Papua New Guinea, Fiji, Solomon Islands, and Vanuatu, access to insurance should not be an issue (Figure 1) as these countries have insurance companies domiciled in these regions, which can provide most forms of construction insurance to those markets. Meanwhile, Cook Islands, Samoa, and Tonga, for example, have some companies available, but they lack the capability to provide insurance to construction projects and would not have the capacity to insure large projects. For smaller and/or more isolated countries like the Marshall Islands, the Federated States of Micronesia, Kiribati, Nauru, Niue, Palau, and Tuvalu, there is no suitable regulation

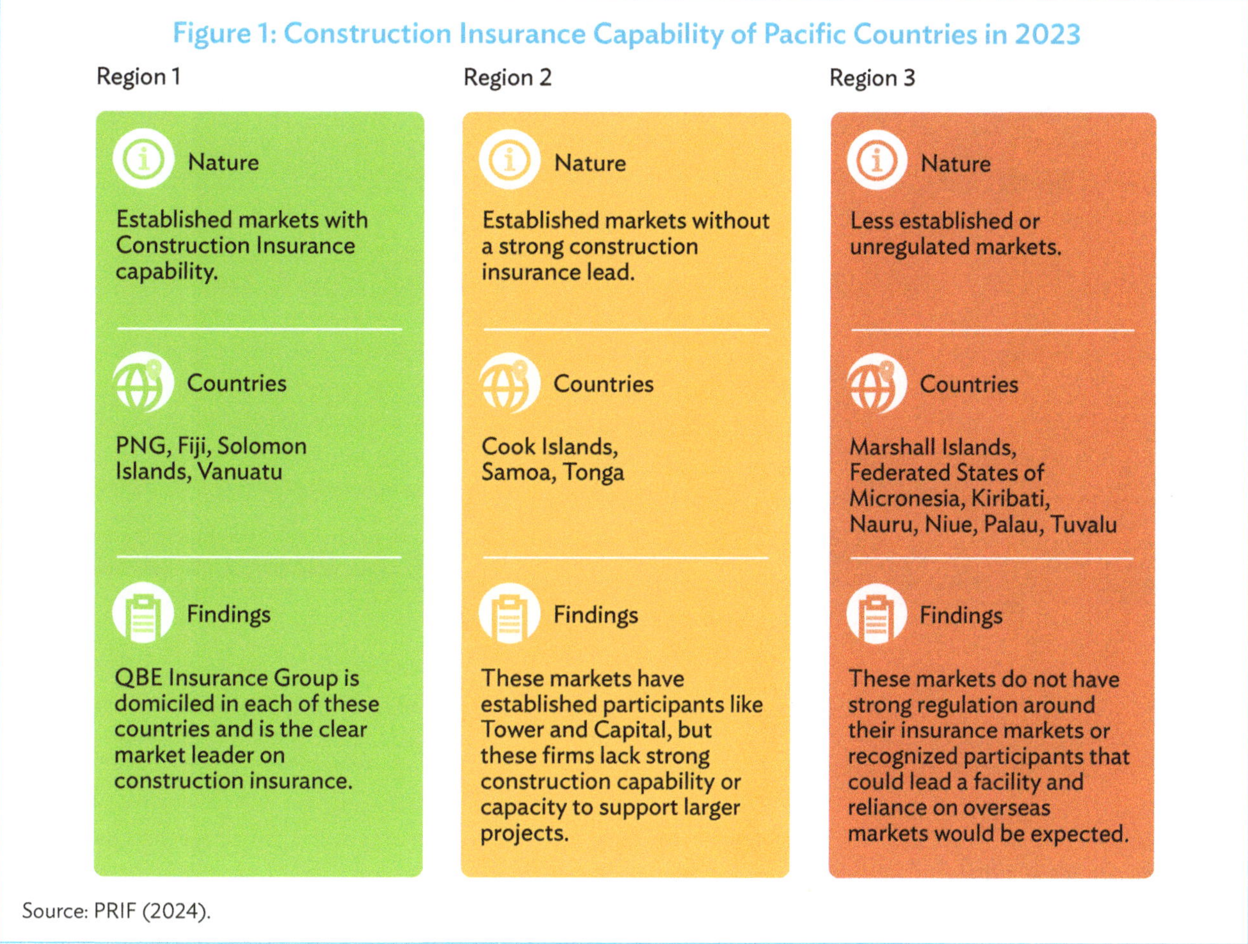

Source: PRIF (2024).

or insurance companies in these countries, leading to full reliance on overseas markets. This likely impacts local firms on small projects who have issues in sourcing insurance from overseas firms, while for international contractors, this will be less of an issue. Notably, this insurance issue is the status quo and may deteriorate further given increasing risks due to climate change and rising sea levels.

The Construction Sector in the Pacific

Although climate change was highlighted during the scoping interviews as a major risk, this report also views it as an opportunity for construction companies. Climate adaptation will generate demand to erect or upgrade dams, build sea defenses, and protect infrastructure and buildings from rising sea levels.

More generally, supply chain risk management is not the direct management of individual parties or companies within the supply chain but is the management of the risks themselves. In the construction sector, supply chain risks are defined as any potential events or issues associated with any entity within the supply chain that, if experienced, would have a negative secondary or   knock-on effect on an organization's objectives (be it in a construction company or Pacific government) (Thompson 2022).

In the Pacific, construction companies can be based on one island or overseas. As such, material, equipment, and labor are often imported. In addition, profit and responsibility centers for those construction projects may be far away from the project sites themselves. Based on gravity modeling, Alexander and Merkert (2017) have shown that distance can be a major obstacle to trade, and it is likely this also applies to the construction sector in the Pacific islands. Scoping interviews with donor organizations such as ADB and the World Bank, as well as with logistics companies and ports, highlight that distance, remoteness (Figure 2), and the lack of connectivity in Pacific Islands are hard to overcome.

Scale and Pacific Exports and Imports

The Pacific economies are small. Tuvalu's exports are one of the smallest of any country listed by the World Trade Organization (World Bank 2023). Total 2022 exports of the 14 Pacific countries in this study, of $17 billion, are only 0.068% of total global exports. Within this figure, exports from Papua New Guinea (PNG) account for almost 90%. While a COVID-19 effect is visible in the data in Appendix 2, many Pacific countries seem to have recovered and are growing again, especially PNG, which highlights its reliance on transport infrastructure.

For merchandise imports (roughly half of the value of exports), the relative importance of the 14 Pacific countries is even smaller, with 0.034% of total global imports, as shown in Appendix 3. Similar to exports, imports were growing again in 2022, with a few Pacific countries, like PNG and Fiji, accounting for most (circa 70%) of the imports. In contrast to exports, other Pacific countries do have notable imports, such as Palau, Solomon Islands, Samoa, Tonga, and Vanuatu. This can be seen as beneficial to—in fact partially due to—the construction sector.

Logistics Performance of Pacific Countries

In the latest World Bank Logistics Performance Index (LPI 2023), only three of the 14 Pacific island countries are included in the ranking: Fiji, PNG, and Solomon Islands. All three countries are not strong relative to the best performing regional economies, such as Japan and Singapore. PNG and Solomon Islands have improved their scores and relative ranking compared to the last round in 2018, while Fiji worsened its score and ranking and is now in the bottom 10% part of all global economies (Table 1).

Table 1: Selected Logistics Performance Index Scores in 2018 and 2023

Economy	2023 LPI Score (out of 5)	% change from 2018	LPI Grouped Rank (out of 139)
Singapore	4.3	7.60	1
Hong Kong, China	4.0	2.00	10
Japan	3.9	(3.30)	14
Australia	3.7	(1.37)	19
New Zealand	3.6	(7.77)	
Solomon Islands	2.8	8.90	73
Papua New Guinea	2.7	24.19	79
Fiji	2.3	(2.20)	123

( ) = negative, LPI = Logistics Performance Index.
Source: World Bank Logistics Performance Index (LPI 2023).

To better understand the observed changes in LPI scores for Fiji, PNG, and Solomon Islands, Figure 2 shows the relative change of the various indicators that together form the LPI score. While the relative customs performance has deteriorated (a supply chain risk in its own right), the infrastructure score has substantially improved for PNG (up 22% from 2018 to 2023) and Solomon Islands (up 18%) but deteriorated for Fiji (down 8.4%). The international shipments score has improved for all three, with Solomon Islands up 32%, followed by PNG and Fiji. PNG has improved the most in both logistics' performance, quality, and timeliness, with timeliness being a real concern for Fiji, which has worsened in this subcategory of the total LPI score. This is a clear risk for any sector depending on international supply chains in and out of Fiji and is further explored and evaluated in the discussion section.

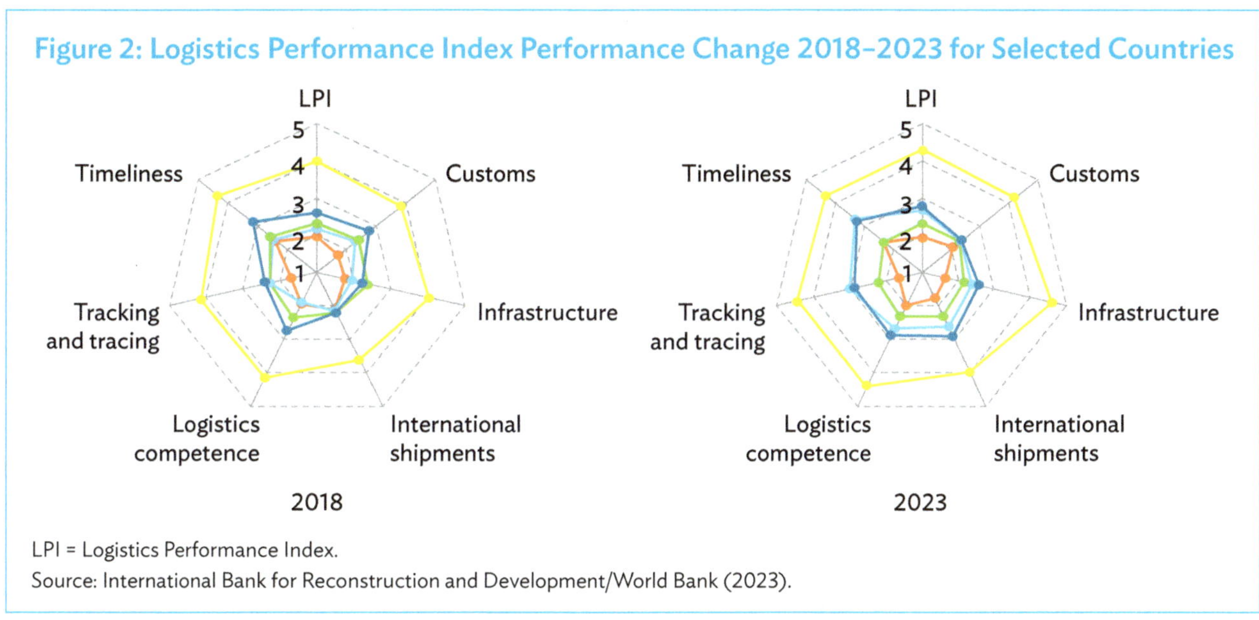

Figure 2: Logistics Performance Index Performance Change 2018–2023 for Selected Countries

LPI = Logistics Performance Index.
Source: International Bank for Reconstruction and Development/World Bank (2023).

The Pacific construction sector has much to do with the development of transport and logistics infrastructure such as ports. For example, as shown in Table 2, the majority of the large value (i.e., > $10 million) construction contracts procured using ADB and World Bank financing, during 2018 to 2022, were related to transport infrastructure (ADB = 67%+, World Bank = 83%+). Table 2 further shows the distribution of awarded contractors across regionally based and international contractors. As such, bidders for donor-procured projects are limited in number and geographic spread, requiring additional efforts to increase the pool of contractors interested in bidding on Pacific business opportunities.

Table 2: Distribution of ADB- and World Bank-Procured Large-Value Construction Contracts 2018–2022

	ADB		World Bank	
	% of total number of contracts	Number of unique companies	% of total number of contracts	Number of unique companies
By Sector				
Sector				
Transport	67.57	. . .	83.33	. . .
Water and urban development	10.81	. . .	16.67	. . .
Energy	18.92	. . .	0.00	. . .
Human and social development	2.70	. . .	0.00	. . .
By (Lead) Bidder Nationality				
People's Republic of China	62.16	15	33.33	1
Pacific island countries	16.22	4	33.33	1
Australia/New Zealand	10.81	3 Aus + 1 NZ	16.67	1
Rest of world	10.81	2 India, 1 France, 1 United States	16.67	1

. . . = not applicable, ADB = Asian Development Bank, Aus = Australia, NZ = New Zealand.

Note: The data relates to unique winners. This may be firms that won a contract as a single entity or as the part of a joint venture. If a firm won a contract as both a single entity and as part of a joint venture, it is counted twice. For the purposes of reporting, where a joint venture won a contract, the nationality of the bidders is assigned to that of the lead joint venture partner.

Source: Based on raw data from the ADB and the World Bank.

The Research and Evidence Gap and How This Study Can Add Value

While this study focuses on investigating supply chain risks in the construction sector in Pacific countries, a subsequent question emerged during the first phase of the project (first literature review and scoping interviews) that has potential to advance knowledge beyond what the World Bank (2023b) reported. The present study, in Appendix 6, therefore explores why there are so few local and international bidders and companies competing for and undertaking donor agency procured construction projects and whether supply chain risks are a main cause of this. We reveal if and how these perceived risks can be overcome, mitigated, or minimized through specific strategies, policies, governance frameworks or tools. We aim to understand what, in the perception of construction contractors and companies, the supply chain risks are and how they can be overcome, with the overarching aim being to increase both the number and diversity of bidders for donor agency construction tenders and projects and reduce associated bid prices through adequate risk management and mitigation.

The World Bank (2023b) discussed several procurement approaches that may help itself and borrowers (including Pacific countries) to increase the number of bidders for its projects. This includes consolidating the pipeline of infrastructure projects to make them more attractive to bidders and offering a sequence of projects in the same location to offset high mobilization costs. Although this may attract more international bidders, it risks making it more difficult for local companies to compete.

The scoping interviews established that both ADB and the World Bank organize regional pipeline and contractor presentations and business opportunities seminars for contractors, run by the Australian Trade and Investment Commission (Austrade) and the New Zealand Trade and Enterprise, but, based on interviews with ADB and World Bank staff, it was stated that attendance is predominantly by consultants. This has the disadvantage of limiting transfer of knowledge between the agencies and contractors. While the legal relationships are between Pacific governments and contractors, there is a lack of direct contact between ADB and the World Bank and both Pacific and international construction companies when it comes to knowledge dissemination about their procurement policies and upcoming business opportunities. This may also limit their knowledge of construction companies' preferences when it comes to bidding on Pacific projects and their perceptions and concerns around supply chain risk and potential measures that could improve their willingness to participate in projects financed by both institutions.

3 Methodology

Interview Program

Data collection was mainly accomplished in an interview program with the private sector, development banks, and government stakeholders from July to October 2023. To close the evidence gap identified in our literature and document review, we reviewed multilateral development bank portfolio profiles for awarded infrastructure contracts to determine likely supply chain impact targets and to identify a representative sample of interview subjects. We conducted exploratory interviews with a representative sample of contractors, including contractors awarded contracts in the Pacific (i.e., under ADB, World Bank, New Zealand's Ministry of Foreign Affairs and Trade, or Australia's Department of Foreign Affairs and Trade projects) and those of reasonable caliber but reluctant to participate in such tenders and construction industry experts. The interviews aimed to identify key supply chain challenges affecting their operations, the import of construction materials into the Pacific, as well as the constraints and drivers for participating in regional procurement opportunities.

The main interview program consisted of semi-structured interviews of 1 hour, on average. The interview guide is in Appendix 4. In addition to mapping unique supply chain challenges in Pacific islands, the interviews included questions around measures to mitigate identified supply chain risks by reducing bottlenecks and supply chain costs, and indicators that may enable government and development partners to consider risk when designing projects. We were also asked to investigate modalities to assess how supply chain risks influence the pricing of infrastructure projects, that is, risk premiums that may be associated with supply chain issues.

Interviewees

In contrast to Al-Mhdawi et al. (2023), who had responses from a large sample of construction contractors, the pool of actual and potential local and international contractors in Pacific countries is substantially smaller. Therefore, in addition to contractors, the interviewee pool for this exercise consisted of:

- Eleven managers from eight active and experienced construction firms operating in the region.
- Four suppliers active in the region.
- Five large contractors and four suppliers not currently active in the region.
- Eight logistics companies, such as local shipping and airlines, as well as ports and airports.
- Five Pacific governments or ministries.
- Five international government authorities interested in promoting business development in the region.
- Seven financiers, including ADB and the World Bank.
- Three academics and consultants from the region, as well as three international engineering consulting firms.

In total, we interviewed 59 managers of 52 entities active in the construction supply chain ecosystem in the Pacific countries. Despite the semi-structured nature of the interviews, through open-ended questions we provided the interviewees with flexibility and freedom to explore supply chain issues and topics they deemed important. Appendix 5 lists interviewed stakeholders.

Participation at Industry Events

To triangulate and confirm our interview data, in the final steps of the primary data collection we participated in two industry events: a half day International Development Opportunities Seminar in Sydney (organized by Austrade) in August 2023, and PRIF Week, one of the largest Pacific infrastructure-focused conferences attended by many Pacific government officials, in October 2023. Both events were well-attended, presenting many opportunities to converse with a multitude of stakeholders and to confirm findings obtained from earlier parts of the data collection.

4 Results

The results cover two main topics:

- development and findings of the framework to map Pacific country supply chain risks and challenges in construction supply chains, and
- risk mitigation strategies for contractors.

Development and Findings of the Framework to Map Pacific Country Supply Chain Risks and Challenges in Construction Supply Chains

Synthesis of General Findings and Risk Matrix

A first finding is that many of the risk events have already occurred and hence manifested as past or current challenges rather than risks. Second, the characteristics and supply chains of each Pacific country are complex and often differ the issues encountered and how much they impact the overall supply chain risk.

We have identified a non-exhaustive list of the key supply chain risks and challenges highlighted by our respondents, placing an increased focus and higher weight on risks identified by both current and potential contractors. The patterns of risks differ across the three main regions of the Pacific countries: Melanesia, Polynesia, and Micronesia, as shown in Table 3. The reason is related to geography (i.e., remoteness and the vast distances both from supply and demand markets and within the Pacific countries) as well as to historical connections to larger countries and markets (Australia, New Zealand, the United States (US), and and the People's Republic of China [PRC]).

Construction supply chains in the Pacific countries involve both bulk (i.e., materials loaded directly onto ships) and container shipping; large container shipping lines are not so interested in Pacific countries or in larger economies such as Australia or New Zealand, due to the lack of economies of scale. The issue is compounded by Pacific country reliance on imports, since they do not have their own manufacturing sectors, and this reliance on other external markets increases their supply chains. In addition to low trade volumes, the Pacific countries are very far away from important source markets (e.g., Japan, the PRC, the US) and the distances to some of the outer islands, or even the 2,704 kilometer distance from PNG to Vanuatu, are substantial.

The added volatility of Pacific supply chains makes them very complex to manage, which can be a substantial disincentive to international contractors entering the Pacific construction market. As one contractor noted: "we see ourselves as a logistics and supply chain management company that happens to be in the construction sector. Our unique selling point and competitive advantage is to have extensive local expertise with logistics

and supply chains in the Pacific islands." This knowledge and expertise, gained through many years of working in the Pacific, is generally not available to new entrants with limited experience of delivering projects outside their own developed countries with associated robust supply chains. Hence, they perceive substantial risks or may underestimate the actual complexity of project delivery.

To systematically cluster the Pacific countries by scale and size in our supply chain context, all relevant countries were ranked based on their merchandise imports in 2022. For Melanesia, all four Pacific countries are shown independently, but, for Polynesia, only the largest two are separated, with Cook Islands, Niue, and Tuvalu grouped together due to their similarities in size and supply chain risks. All Micronesian islands in the Pacific were grouped together for the same reason. The study was further complicated by the focus of many contractors on particular islands, leaving them without a comprehensive view of risks and challenges across all the Pacific countries.

Table 3 orders the challenges and risks by their importance and frequency of mention during the interviews.

Table 3: Pacific Country Construction Supply Chain Challenges and Risks Matrix

Challenges/risks	Melanesia						Polynesia	
	Papua New Guinea	Fiji	Solomon Islands	Vanuatu	Samoa	Tonga	Cook Islands, Niue, Tuvalu	Marshall Islands, Federated States of Micronesia, Kiribati, Nauru, Palau
Merchandise imports in 2022 ($ million)	3,189	2,984	644	413	435	197	174 (=124+16+34)	669 (=223+106+35+211+94)
Cost of shipping	low	low	low	low	low	low	low	low
Shipping service supply chain risks[a]	low	low	medium	medium	medium	medium	high	high
Lack of port infrastructure[a] (incl. cranes)	low	medium	medium	medium	medium	medium	high	high
Air cargo service supply chain risk	medium	low	medium	medium	medium	medium	medium	medium
Lack of cargo airport infrastructure	low	low	medium	low	low	medium	medium	medium
Lack of local construction material supply	low	low	medium	medium	medium	high	high	high
Lack of local construction equipment	low	low	medium	medium	medium	medium	high	high
Skill shortage risk	medium	medium	medium	medium	medium	medium	high	high
Staff attraction and retention	medium	medium	medium	medium	medium	medium	high	high
Foreign exchange risk	high	medium	medium	medium	medium	medium	medium	medium
Supplier risks[b]	medium	medium	medium	medium	medium	medium	medium	medium
Supply chain visibility	medium	medium	medium	medium	medium	medium	medium	medium
Health/access to water etc. risk	medium	medium	medium	medium	medium	medium	medium	medium
Change of local regulation/enforcement risk	medium	medium	medium	medium	medium	medium	medium	medium
Local government capability risk	medium	medium	medium	medium	medium	medium	medium	medium
Unrealistic expectations of clients (PIC gov)	medium	medium	medium	medium	medium	medium	medium	medium
Inflexibility of ADB/WB procurement/contracts	medium	medium	medium	medium	medium	medium	medium	medium
Payment delay risk	medium	medium	medium	medium	medium	medium	medium	medium
Short-term climate change	low	low	low	low	low	low	low	low
Long-term climate change	medium	medium	medium	medium	medium	medium	high	high
Insurance risk	low	low	low	low	low	low	high	high
WHS/ESG risks								
COVID-19/exogenous shock risk	high	high	high	high	high	high	high	high

● = low risk, ● = medium risk ● = high risk

COVID-19 = coronavirus disease, ESG = environmental, social, and governance, gov = government, PIC = Pacific island country, WHS = workplace health and safety.

[a] Connectivity/frequency, delays, reliability, and cancellation.

[b] Delays and quality of procured construction materials.

Note: As Reported by stakeholders interviewed for this study.

Source: Asian Development Bank.

Shipping Issues

Extremely high shipping costs and associated prices for loading, packing, and all other logistics services were a major concern during COVID-19 and showed that the Pacific countries were heavily exposed to this global shock, with many ripple effects exacerbating local issues. Shipping prices have normalized since then and, as of December 2023, are below pre-COVID-19 levels. Only a few contractors voiced concerns about lagged effects, which are no longer seen as a challenge or risk. There are still regional differences between shipping lanes (prices, frequencies, and distances), which reportedly puts Australian and New Zealand exporters at a disadvantage compared to Malaysia or other Asian countries.

During the pandemic, shipping services were severely impacted with, for example, only one in five ships arriving on time, frequent changes of rotations and, if not enough cargo, ships choosing not to stop at particular ports, instead docking during the next rotation. As of December 2023, this situation has substantially normalized. Many reliability issues (connectivity, capacity, frequency) remain, however, and are often compounded by inappropriate wharf infrastructure. This is especially acute for outer islands and the more remote and small islands of Micronesia, which need landing craft, because they do not have port berths or barges. Likewise, in Polynesia, weather can prevent discharging of vessels for several days. The supply chains of remote islands are not only longer and at the very end of global supply chains, but shipping lines also typically treat them as the lowest priority, disrupting them more. Mid-sized islands (Polynesia) fare slightly better but also suffer wharf infrastructure and weather issues. Expensive specialized equipment is required to get materials to these remote islands. And inter and intra island shipping is an issue, with only two or three firms operating with vessel size and frequency limitations which makes pricing difficult. Reportedly, those shipping fleets are also not well maintained, which leaves those services unpredictable. Interestingly, Solomon Islands has built strong supply chain links with northeast Asia and operators from that region are happy with shipping service quality to that Pacific country. However, this view is not uniform among all stakeholders interviewed, particularly those with supply chains emanating from other regions.

Vessel size limitations mean that vessels are optimized for certain purposes (and cargo) and operators may be overwhelmed when new projects are awarded that require either bigger or smaller vessel dimensions. While conventional and containerized freight appears to be less affected—although empty container movements are an issue—construction projects usually require break-bulk cargo, which is problematic as carriers have capability and capacity limitations and port infrastructure weight and dimensions are limited on most islands. Damage to oversize cargo in transit from rough seas is another risk. Inconsistent dimensions and weight of construction materials also makes packing efficiently and safely a challenge.

Planning for transport in construction supply chains during contract bidding is difficult as contractors need to estimate and factor in conditions and prices several months ahead of shipping. Extreme weather in the October–April cyclone season also plays a role when, for example, a project in Fiji may be fine, as Fiji is a shipping hub, but other islands might be skipped due to inclement weather. Tonga and Samoa are not on the main routes and get calls more infrequently. Even larger islands suffer from this, as ships that sail toward Pacific countries through New Zealand's Port of Auckland transshipment hub do not go to PNG. Factoring these considerations into bidding durations, such as providing more than the existing standard 6-weeks for Pacific projects, would provide existing and potential contractors additional time to plan and cost logistics in their bids.

Even in the larger Pacific countries, even if green in Table 3, interviewees reported substantial delays in offloading and clearing cargo. Issues included regional port infrastructure worse than the main port of the country, ports focusing on non-construction commodities (such as palm oil), or regularly congested ports (e.g., the Port of Suva in Fiji). Since COVID-19, demand for cargo has increased, exposing a lack of shipping capacity. More recently shipping lines or companies such as Swire Shipping and Mediterranean Shipping Company have added services that could cause bottlenecks in port infrastructure.

Another issue in the supply chain risk is that some Pacific countries do not allow international or foreign shipping lines to berth and the number of competitors for shipping services is very low: Tuvalu has only one shipping line, while most other islands have two. Developing a better understanding of the specific shipping constraints that may impact a project in planning could reduce downstream implementation-related risks.

In Fiji, Neptune Pacific Direct Line—an important local shipping line and sister company to Fiji Waters—does not always have capacity for construction material, as its main business objective is exporting water from Fiji to the US and the PRC. Talking to Neptune Pacific Direct Line revealed its provision of essential supply line services to (i) Australia/New Zealand–Fiji—Samoa–Tonga, (ii) New Zealand–Vanuatu and back via Fiji, (iii) New Zealand–Tahiti, (iv) break-bulk services into Tonga, and (v) charters into PNG, Tuvalu, and Cook Islands. Nauru now runs its own single vessel shipping company with substantial risk and Neptune Line regularly having to help out. In addition to the liner services, Neptune's other ships are geared to discharge in different terrain and are chartered by requests from ADB, the World Bank, or contractors to move large equipment or even houses that would not fit on a container ship.

Overall, shipping service availability and reliability is seen as a major supply chain risk and the inconsistency of time schedules of shipping lines reportedly affects building programs (especially as shown in red in Table 3). Pricing of that risk is difficult and attracts a premium or risk margin that may drive contractors out or bar entry of newcomers.

As part of cost considerations, time is vital in Pacific country supply chains, not only from an on-time or delay perspective, but also due to often poor reliability and inconsistency of transport chains. For contractors to deliver on time, they need to ship and occasionally fly labor and materials to an island, including fuel. Given that a large proportion of electricity on the islands is still generated by diesel, this is a big issue, both for supply chain resilience and environmentally. Several interviewees pointed out that Fiji's biggest foreign exchange expenditure is diesel. Government officials noted that Fiji is now becoming advanced in producing sustainable aviation fuel, but airlines noted that it remains very expensive and not the best solution when the rest of the economy is running on diesel. This is even worse in the outer islands, where diesel for energy needs to be shipped over long distances.

Demand exists for air freight, meanwhile, but it is typically minimal, as it is on average 10 times more expensive than ocean shipping. Apart from regular air services to/from PNG and Fiji, only three dedicated freighter aircraft offer large craft in the region. People rely instead mainly on B737 passenger aircraft "belly-hold" capacity. Infrastructure is also lacking on some smaller and remote islands. There is also a shortage of pilots and engineers to fly and maintain aircraft fleets, which is a risk for freight but also for fly-in-fly-out construction labor.

Construction Materials

Interviewees said that risks associated with specific construction materials depend on the specific project. For example, bridge projects, which are relatively common in this region, are generally highly specified and need high quality cement that cannot be sourced from local industry. It must be imported from overseas and then attracts large duties (e.g., 30% in PNG), as does reinforcing steel bars, and therefore is costly. Instability and uncertainty related to frequent tax and tariff changes are seen as a substantial supply chain risk. This is exacerbated by cash flow issues, as this is a long supply chain (4–6 months), which interviewees said is a particular problem in PNG. Construction materials such as cement, steel, and even sand, are common and crucial ingredients to deliver infrastructure on some islands. How to best do this logistically and how Pacific countries may be able to support each other in this regard is an area worth further study.

Labor Issues

Interviewees firmly placed labor issues as a key risk in successfully delivering construction projects in Pacific countries. There is a substantial shortage of skilled local labor (although this varies across the region) and skilled international staff willing to travel to the Pacific islands to deliver projects (due to concerns about housing, health care, and cost of living, among other issues). While it can be expected that it is difficult to attract and retain qualified staff to work on outer and small islands, such as Tuvalu, due to their remoteness, isolation, and lack of access to basic services, it is reportedly equally hard to attract staff to larger islands, such as PNG. Set-up costs and efforts in understanding the market are prohibitive, hindering market entry. Some contractors highlighted that getting visas is often extremely difficult, requiring instead that staff travel in and out on tourist visas. In one Pacific country, long-term visas cost $6,000 per person and decisions are often delayed to the last minute, substantially impacting the contractor's ability to plan contract implementation, it was reported.

Another critical issue is people management. Contractors claim that if staff and management are of good quality, the project should be fine—expertise matters. However, contractors have difficulty attracting high-quality staff, as they prefer to stay in Australia and New Zealand, or other neighboring countries, such as the Philippines or Japan. Compounding this issue, skilled people from Pacific countries get recruited to Australia and New Zealand, often outside their field of expertise, such as civil engineering, as lower skilled jobs like picking fruit or other labor programs in Australia pay them a higher salary.

Foreign Exchange Rate Volatility

Foreign exchange rate volatility (rapid devaluation) was commonly quoted as an important supply risk to be considered. For example, exchange rate volatility in PNG in 2023, combined with the difficulty of locals and businesses to obtain foreign currency, meant suppliers could not do business, as even purchasing fuel (often traded in US dollar) was difficult. Local currency pegs to other international currencies complicates foreign exchange management, such as the Fijian dollar peg to the US dollar. At face value it may appear to be a low risk, as most large construction projects in the region are paid in US dollars. However, many contractors are from developing countries and often the currency of payment is not pegged to their "home country" currencies and many materials come from overseas. Any change in foreign exchange rates therefore can have substantial impact on revenues and costs, and the overall profitability of projects. This risk is somewhat mitigated (for contractors) on most MDB-financed projects, which allow them to be paid in multiple currencies based on the origin of the contract inputs. It appears, however, that this is not known to all contractors and even less so by potential contractors. It was further stated in several of the interviews that the International Monetary Fund is proposing that the government devalue the local currency of PNG, which creates substantial procurement risks, living standard uncertainties, and supply chain risks.

Local Governance

Each Pacific island country has its sovereign regulations, with different health and safety compliance and regulations in each country, varied tax regimes and laws. Existing contractors, especially potential contractors, mentioned governance weaknesses as a key area of concern. Some contractors reported significant delays receiving visa approvals for workers and customs clearances for goods and equipment arriving at ports in Pacific countries, such as PNG and Solomon Islands, undermining their ability to efficiently deliver contracts. This was perceived as providing an opportunity to less principled contractors to accelerate approvals of their permits and approvals through unofficial channels, making the playing field uneven for all bidders and contractors. To encourage firms to do business in the Pacific, donors are expected to set clear expectations with host governments around compliance with established laws and supporting efficient delivery of projects through expedited approvals and clearances.

Transparency and Perceived Lack of Level Playing Field

While it is not directly a supply chain risk, a perceived inflexibility of the funding and contracting model was frequently highlighted as a core issue for interviewed national and international contractors competing for MDB-funded construction projects in the Pacific. Contractors see the model as overly cost focused and not reflective of risks, especially in outer islands. While there was some acknowledgement that the approach has improved over the years, respondents said that a further overhaul and more collaborative approach to sharing risks is needed. As local governments are the clients, contractors also complained about corrupt and non-transparent business practices on projects of interest. These practices place international, publicly listed companies who refuse to engage at a disadvantage, they said. Contractors further reported that they often missed out on projects where competitors offered inferior quality, with World Bank and ADB bid evaluation methods that awarded contracts to technically compliant bidders with the lowest price seen as the main contributing reason. However, this risk is not specific to MDB projects and, in practice, those projects are less risky than when Pacific governments conduct their own projects. This is because the MDBs require additional safeguards in bidding documents and contracts and conduct regular oversight and missions to identify illicit practices.

Some contractors complained that evaluating the performance of contractors, post construction, is not rigorous enough and that underperformers are not penalized (e.g., barred from tendering for future projects) as there are often other (sometimes political) considerations taken into account. Donor contracting procedures need to also better recognize that some international contractors (e.g., Australian and New Zealand companies) need to comply with labor and work, health, and safety regulations of their domestic markets and this adds cost premiums to jobs compared to some competitors from less rigorous jurisdictions. Safety standards in Australia and New Zealand, where many contractors are based, are very high compared to the Pacific islands. Compounding this is that Australian, New Zealand, and international contractors from France, India, Japan, and so on, have tax obligations in their home jurisdictions (behind the scenes costs) that they need to price into their bids.

Current and potential contractors strongly perceive that the Pacific country construction sector is not a level playing field, with certain international contractors having a competitive advantage, particularly when bid evaluation methods focus on price. Contracts are then awarded to these entities, and many believe quality suffers as a result, although some engineering consultants countered that they had not seen sufficient evidence of such practices. Regional contractors also highlighted that local economic development from the work is often limited when contractors bring in their own labor, camps, food, equipment, and construction materials. International suppliers also feel they are being practically excluded from many projects when incumbent contractors are already present in the country, given the large mobilization costs in establishing a presence in Pacific countries. Some interviewed suppliers who managed to win projects in joint bids with large international contractors alleged that they were then kicked off the project as soon as it had been awarded. Several contractors highlighted in the interviews that they lobbied the MDB and their respective home governments to not award contracts to contractors with the lowest cost bids. That said, MDB-funded projects have standard provisions to allow governments to exclude abnormally low bids and include provisions for quality-based evaluation methods, both of which can mitigate the concerns raised.

Climate Change

Clients and the government agencies noted climate change impacts, but this was not seen as an immediate risk to contractors. Climate change and sustainability are seen as a long-term risks that do not impact day-to-day business in individual construction companies or suppliers. According to one interviewee, the priority is "to get the job done and to survive the next month." This likely also reflects the relatively short-term nature of construction projects.

None of the interviewees indicated climate change as a top-10 supply chain risk, contrary to what we expected at the outset of data collection. Contractors explained that having been exposed to severe weather events such as hurricanes and floods, the related risks are manageable, requiring good management of logistics and supply chains. In fact, interviewees mentioned multiple times that climate change would actually be a source of future construction demand in the region, with climate mitigation and adaptation contracts strengthening the project pipeline in the region. As such, climate change was frequently viewed as an opportunity for construction, not just for decarbonization and affordable clean energy (e.g., solar panels), but also other UN Sustainable Development Goals, such as in reducing poverty and hunger, improving education, in gender equality, and in the development of industry, innovation, and infrastructure. Inexperienced potential contractors were more concerned about the impacts of climate change (e.g., hurricanes and floods) on construction supply chains and project viability in the Pacific countries.

Politically, environmental sustainability is high on the agenda, with Fiji a particular champion of the cause and wanting to build sustainable buildings. Interviewees also acknowledged that extreme weather events, such as cyclones and floods, have always been a risk factor in the region. And even though such events seem to have become more frequent and severe, many respondents felt nonetheless that extreme weather events were localized and of short duration. Climate change is also reportedly changing the construction season, with one interviewee noting the Solomon Islands as an example, as measured by wet weather days and high-water levels. However, contracting authorities see these issues as force majeure. The generally accepted view is that holding contractors accountable for this would not be fair (e.g., during COVID-19 delays and cyclones) and authorities are not enforcing clauses (e.g., insurance, delays, force majeure). Authorities also fear they would be left with no contractors if enforced, or with projects that cannot be completed.

Insurance

What is seen as a major issue due to climate change is that some Pacific country contractors cannot obtain insurance, as discussed earlier and in PRIF (2024). While large contractors can access insurance on the international markets and may also self-insure their portfolios, smaller, predominantly local contractors cannot. Because most multilateral and bilateral financing agencies insist on insurance, this can become problematic and limit competition. Reportedly, for local contractors, the risk is small enough to forego the insurance for the most part (if possible and if not financed by MDBs, as otherwise they would be in breach of contract) and for governments to self-insure those small works. This matter is subject to a current and parallel research study by ADB in partnership with the World Bank.

Other Issues

A myriad of "smaller" issues are relevant. As just one example, two interviewees pointed out that telecommunications and internet services are still inconsistent across the Pacific and, the more remote the island, the worse it gets. Reportedly, most Pacific countries advise use of locked-in (proprietary) systems contracted through those countries and to use those as a Pacific country revenue generator. For instance, Starlink is not permitted in some Pacific countries.

Experience helps manage supply chains efficiently, interviewees noted, as does leaving equipment on the more remote islands for use on future projects. While these aspects benefit incumbents, they can also represent barriers to market entry for interested new entrants due to the lower mobilization costs of the incumbents.

Short-term and Long-term Outlook

Interviewees highlighted that the lack of understanding of how Pacific islands operate can substantially increase supply chain-related risks. However, once this local know-how is acquired, they believe it can be used as a unique selling point and competitive advantage while managing the related risks. Notably, local exposure to Pacific country-related risks is different from global exposure. For contractors, risk exposure to Pacific country projects varied from 2% to 100% of their total construction revenues. As such, risk sharing across project portfolios is an advantage enjoyed by large, multinational contractors.

The short-term (1–2 year) outlook for construction project supply chain risks in the Pacific islands is consistently positive with all respondents predicting supply chains to be smooth in the near future. Minor concerns are that some of the shipping capacity lost during COVID-19 has not been replaced, which will lead to further capacity constraints, as there is so much work in the pipeline.

The mid-term (3–10 years) outlook is less clear, with many interviewees concerned about geopolitical tensions, trade routes becoming blocked, insurance becoming less available and more expensive, and an exit of shipping lines more likely than entry. The insufficient labor market is also likely to persist. Larger islands, such as PNG and Fiji, have enough people and many immigrants, and thus upskilling is possible. However, other islands, without that human capital, will struggle to attract and retain people and local contractors to find skilled workers to deliver high quality projects. International contractors will continue to find it difficult to attract skilled workers from overseas to travel to these countries for project implementation. The foreign exchange issue in PNG is predicted to get worse and is a major risk for trading businesses and investment companies, with some construction projects being 2 years late, with firms unable to get materials, due to cash flow and market availability. Interviewees also noted a lack of visibility of future project pipelines, limiting ability to plan for work over the medium term.

In the long term (10 years or more), risks include climate change, including the need to plan construction schedules around more extreme weather, geopolitical risks, and other global uncertainties. Contractors were concerned about whether they would still be able to source supplies due to scarcity and circumstances becoming more volatile.

On a more positive note, many interviewees referenced the fast rate of development in the Pacific, with the provision of internet services and food supply likely to improve further, mainly due to government and donor-funded projects. In addition to rising sea levels, they were concerned about extended droughts, poor health care, water supply, education, and waste disposal. The 2030 net zero greenhouse gas targets will require development. Also an issue, Brisbane's hosting of the 2032 Olympics for Australia will divert many construction resources and skills to the Brisbane area, which could further complicate labor issues in construction supply chains in the Pacific.

Potential contractors and suppliers can currently find many risk-free projects in Australia, New Zealand, Japan, and other developed markets and are not immediately attracted to the Pacific. We identified potential interest from Japanese and German companies. The Japan International Cooperation Agency is a useful intermediary for Japanese companies and Germany has just opened an embassy in Fiji, which is also useful for such efforts. However, while potential contractors recognize the high reward that may come with branching out to the Pacific, when risks (including political and corruption) are considered, they believe they get similar returns at larger scale in safer jurisdictions. As such, they first want to set up subsidiaries and/or grow existing activities in Australia and New Zealand (for some international players this region is still a blank spot) before branching out to the Pacific, which ultimately turns them into something similar to their Australia and New Zealand competitors.

It is worth nothing that many companies prefer to collaborate, at least in the first instance, with large and experienced contractors who have worked in the region before attempting projects independently. This makes the contributions of current contractors and suppliers even more important to the development in this region and both stakeholder groups are seeking support and a cooperative and trusted approach with governments and MDBs to improve the supply chain situation and construction projects in the Pacific countries.

Using our risk framework, we summarize that many substantial supply chain risks exist in construction projects in Pacific countries, many of them not under the control of the contractors and often not easy to build into bids. The vulnerability of contractors is hence high, and the local exposure is very high too. However, the exposure for contractors depends on how much of their portfolio is focused on Pacific countries. Perhaps more importantly, most of the risks and challenges are known and to a degree can be quantified. In addition, the preparedness of contractors and suppliers is very high and expertise with handling supply chains in the Pacific countries context is fundamental. Newcomers have tried and failed. It is a common view of interviewees that for contractors (construction companies) and suppliers that have experience with conducting construction projects in this region, these substantial supply chain risks are navigable. Indeed, due to mitigation approaches, they are manageable and typically financially, socially, and/or environmentally rewarding, depending on the development goal of the project.

Several risk mitigation strategies have been identified that can help overcome and mitigate the risks, some of which are further discussed in the next section.

Risk Mitigation Strategies of Contractors

As discussed, the level of preparedness, institutional capacity, and adaptability in contractors is substantial. Contractors hedge foreign exchange rates and fuel where possible, cash flow permitting. They try to plan at least 3 months ahead and to get large orders through stores and trade bases (i.e., Fiji). But getting supplies to smaller and outer islands is much more difficult. In that context, barges (owned or rented by contractors) are often utilized, as they are typically more reliable, although more expensive.

Shipping Services: Contracts and Commissions

Pacific governments have their own risk mitigation strategies for shipping, such as signing exclusivity contracts with shippers to make them call at their specific island, such as the Matson shipping line to Niue. An interesting government approach to shipping challenges was presented in the case of the Micronesian shipping commission for the three US Compact[3] countries—the Marshall Islands, the Federated States of Micronesia, and Palau—which ensures breakeven operations for shippers. Any carrier who wants to provide services, after winning a bid, rather than hauling into each port separately, uses a coordinated approach aligned with the US that gathers cargo in shipping hubs. This "hubbing" to outer islands is seen as the future, as is protecting routes through a shipping commission and grant of capacity and access to ports for individual shipping companies.

Contractors often find it difficult to take on shipping risks, as they have no control over them. Mitigation is difficult and pricing appropriately would often price them out of contention in the relevant tender. As such, to minimize risk and allow implementation of projects, some contractors often organize their own shipping and freight, especially in maritime construction and for specialist projects for smaller islands. Most contractors look at the lowest price of local shipping lines, but often have their own tugging barges for concrete units. For example, materials are transshipped to Fiji and then moved to Niue with their own barge. Since the COVID-19 crisis, many contractors are investigating options of shipping or barging bulk materials themselves, but it comes down to cost and scale, where, especially for civil type projects from $100 million–$150 million, it would make more sense to vertically integrate, as discussed below. But many contractors still rely on existing infrastructure and containerized transport or partner with contractors that are vertically integrated on the shipping side of the business.

---

[3] The Compacts of Free Association are international agreements governing the relationships between the US and the governments of the Marshall Islands, the Federated States of Micronesia, and Palau.

In practice, many different avenues lead to lower shipping risk, including ownership of barges and tow tugs. But those vessels then need to be kept fully utilized, which is not easy, as they are often project-based. Many contractors typically aim to understand, as a first step, what cargo is moved and what contract options are available. For example, in Tuvalu contractors need to move cargo with ship gear and not containers, and break-bulk cargo is beached on islands. This carries substantial risk, as weather regularly prevents beach landings, and the barge may sit there for weeks. In those instances, it is good to own the barge, but the client (government) and donor banks are then still involved to recognize and share that risk. The solution is often a standdown rate, meaning if the weather is not suitable for discharging, there is a payment and/or time extension. This type of risk and cost allocation may lead to a cheaper contract for donor banks and clients in good weather, with an example contract mentioned by one of the interviewees in Tokelau, north of Samoa.

Vertical Integration

Some contractors limit the supply chain risk by vertically integrating with a supplier, for example to get product into the Fiji market, such as manufacturing building products, gip board, steel mills, cement, insulation, and pipes and plumbing. Some take the vertical integration strategy even further, as they prefer vertically integrated services and "survive by engaging expertise" on a self-perform model basis and backing themselves by doing almost everything in-house or with very trusted partners (only one subcontractor over the last three projects). This does lead to minimal subcontracting and so donors like to support local employment, these contractors point out that this is still in that model, as one contractor following that model requires 70% of staff to be local. They raise skills among locals, provide meals at the site, and procure many goods locally, finding that the approach retains staff productivity and supports the local economy. Such vertical integration is good as a risk mitigation strategy but makes market entry for newcomers even more difficult as they then do not have access to local resources which take time to build. Specifically for construction, our interviewees reported that vertical supply chains limit competition in particular for concrete and aggregate (rocks).

Contractors from the PRC are often perceived as 100% vertically integrated, meaning they not only import all construction materials for their projects from the country but bring in their own camps, food, and people and as such do not contribute to local economic development. This approach has also reportedly led to the importation of poor-quality materials at times. It is important that contracts have detailed technical specifications that require working in the upper end of quality. Construction supervision activities should also ensure that contractors adhere to these specifications and that all work is inspected prior to acceptance to avoid a scenario where problems emerge during the lifetime of the asset. Government enforcement of quality can be insufficient, giving cost-leaders a possibly unhealthy advantage when bidding for projects that are awarded solely based on price. Most interviewed contractors, consultants, and potential contractors shared this view. Some Australian and New Zealand contractors have agents in the PRC, which then also help load materials into containers but otherwise contribute substantially to Pacific economic development by using local staff and local materials where possible.

Non-PRC carriers agreed that cost pressures need to be passed on to the end user and vertical integration enables this along the supply chain.

According to the interviewees, one cost effective way of mitigating risk is to reduce the scope of projects to focus on "must haves" rather than "nice to haves," given the Pacific construction project context. While all islands are different, they face common issues, as their supply chains are risky, contractors often keep their own equipment in various countries (vertical integration light) and islands to avoid relying on subcontractors. There are sometimes restrictions on contractors transporting their equipment to outer islands, requiring them instead to own, not lease, plant and equipment, which typically raises capital costs.

Insurance

Another tool for reducing risk is freight insurance for material in transit, which has become popular since the 2018 Suva harbor incident, where a vessel sank and the $800,000 loss was avoided or recovered through insurance. However, cancellation and lateness of shipping services is not covered, which contractors consider problematic. Further research is required in this area.

Labor Training

A risk mitigation strategy interviewed contractors use to reduce uncertainty and labor related risk is to broaden and deepen the experience of people they employ in Pacific associated supply chains. They train their local and international staff and educate consultants to understand the complexity of Pacific shipping and the volatile market. For example, the process to clear a container can change overnight in some ports and little documentation is available.

Management Systems

Another proven risk mitigation strategy to improve forecasting, planning, and operations management is the introduction of management systems that improve visibility of the source of construction project supply chain components. Digitization of that supply chain is in progress in most but not all contractors and logistics companies. Some contractors also mentioned that while technology enables enhanced human rights, such as antislavery or no child labor, as well as robust, transparent, and verifiable (construction and energy) supply chains (where there are materials), it also increases costs of these materials, which then cannot be priced into contracts, unless the evaluation of those contracts place a suitably higher weight on quality over cost.

Foreign Exchange

Foreign exchange risks related to importing materials, such as tiles from Europe or cement from Australia, are typically at least partially mitigated by fixing all procurement costs as soon as a contract is awarded. Some contractors also use financial hedging, yet generally prefer contracts with advance payments and then fixing prices at current rates. As it is not unusual for it to take 6–9 months for a tender to be evaluated in the Pacific, price adjustment matrices included in contracts help minimize risks.

"Cherry Picking" Projects

Another increasingly popular risk mitigation approach contractors use is "cherry picking." That is, they choose not to engage in tendered construction contracts that in their view are designed to be awarded to the lowest cost bidder. Given the construction activity and skill shortages in their own jurisdictions, contractors do not need to take on too much risk by branching out into the Pacific and only do so if they see sufficient reward for taking on the supply chain risks associated with construction projects in this region.

Education of Clients About Risks

Finally, many contractors try to conduct early procurement and educate clients about the supply chain risks of Pacific construction projects. For example, to import 2,000 tons of specialist bitumen for a specific construction project, special tanks had to be organized—temperature-controlled containers—that could be heated to 160 degrees once they arrive on site. The contractor procured them from a trader in Switzerland and the material was then shipped out of Poland via Fiji to Niue. This took considerable time and involved upfront cost and foreign exchange cost fluctuation, as bitumen is traded in US dollars as well as shipping risks and cost variability. The risk mitigation strategies used in this example was to ask the client to accept a bid price in US dollars, include an allowance for price variation between contract signing and order placement and incorporate an advance payment to support some of the contractor's upfront costs.

In Pacific construction projects, engineering consultants are frequently brought in early to discuss specifications and project requirements with all stakeholders to set expectations, to build trust and to identify capable contractors. Much of this is associated with MDB-funded projects. Appendix 6 provides more detail on this matter.

Finally, it was suggested that contractors should be able to use more aggregate construction materials sourced locally in the Pacific countries, and bidding agencies should be more flexible in contract procurement to achieve fit-for-purpose performance specifications.

Feedback from Industry Events

Our attendance at the International Development Opportunities Seminar in Sydney organized by Austrade in August 2023 revealed that there is substantial interest in project opportunities in Pacific countries. The event was sold out and the various MDBs presented healthy indicative 2024–2025 project pipelines, with the World Bank having a pipeline of 39 projects and the ADB of 22 projects. A key focus of these projects is transport infrastructure, and the size of each project is relatively large. While there was a lot of interest during the event, it was notable that much of it was from consultants rather than contractors, which was reported by development partners as being consistent with other similar events held to promote business opportunities in the region. We conclude that this could result in insufficient competition for many of the presented projects.

5 Findings, Recommendations, and Further Research Opportunities

Key Findings

Our findings show that construction projects in the Pacific countries are associated with many supply chain risks, although many appear to have already eventuated and can be classified as challenges, while others have scope for risk mitigation and, therefore positive and attractive project outcomes. As anticipated, the more remote and the smaller the island and market, the higher the risks. As established through our developed framework and evidenced in the findings, there is a pattern but also heterogeneity across the islands, and sometimes even villages on the same island, in terms of culture, governance capacity, access to health services, housing and fresh water, and important transport infrastructure such as airports, ports, and berths.

Climate change has been part of the risk assessment for some time. While it is a substantial long-term risk and extreme weather events are disrupting operations more frequently, in the short term it is perceived as a normal challenge for operations management that contractors are happy to accept. In fact, contractors feel that climate adaptation projects are driving many construction opportunities in this region. Pacific country governments are very a much aware of this and there was talk about adaptation projects in our interviews. Collaboration with Pacific countries in this context but also for shipping commissions is important, as in the example of Micronesia.

The biggest short-term supply chain issue is shipping service reliability, but a myriad of other challenges include a lack of infrastructure, governance, skills and labor shortages, health services, and freshwater access. The biggest single issue in the climate change debate, also relevant for alternative aviation or shipping fuels, is that almost all electricity on many of the Pacific countries is still generated by diesel. This presents not only a supply chain risk to construction projects but a climate change concern that climate adaptation should be prioritized. Many MDB-funded projects aim to address climate change and build resilience, but it will take time.

COVID-19 was a substantial additional shock to this ecosystem. Although shipping capacity has not fully returned to pre-pandemic levels on all routes, supply chains and freight rates have normalized to below pre-COVID-19 levels. Contractors perceive the management of construction projects in remote islands (i.e., the smaller and outer islands) in Pacific countries to be all about logistics and supply chain management. As such, Pacific country construction projects are, by definition, seen as risky and will become even riskier in the future. However, our interviewees see these risks as manageable when contractors have the required expertise, and can ultimately be rewarding financially, socially, and/or environmentally. Many of the challenges, in particular around shipping, are less of a concern in the larger Pacific countries of Melanesia and Polynesia but are more confined to the smaller, more remote, outer islands.

Many interviewees believed that a level playing field is lacking, with a prevalence of state-owned contractors bidding at low prices with a consequent quality deficit, allied with very limited secondary economic benefits arising from the infrastructure delivery. Differences in regulation and foreign exchange rate risk compound this situation. These views were countered by others who noted that there was no distinct correlation between delivery quality and contractor origin while MDBs noted that price is no longer the defining component in the award of large infrastructure contracts.

Governance issues in some Pacific countries and a lack of local government capabilities (with Pacific country governments the clients for most projects in the region) are seen as other important risks, as are certain procurement, monitoring, and enforcement processes of MDBs.

Recommendations

Short-term Actions

(i)   Combine and jointly promote the project pipelines of the major development partners in the region to provide visibility to both incumbent and potential Pacific contractors on upcoming projects in the Pacific countries over a short to medium-term time horizon.

(ii)   Analyze, classify, and present to the contracting market an overview of the various shipping constraints on the main Pacific countries or indicate that this should be undertaken early in the preparation of projects. This would assist clients in understanding potential risk premiums and help contractors with pricing risk.

(iii)   Where feasible, develop engineering designs cognizant of any constraints that may exist in shipping materials and equipment to the project worksite. As a further step, optimize designs to maximize transport in containerized units and reduce break bulk cargo to reduce complexity and cost of transport.

(iv)   Fast-track visas. Given that it appears to be often time consuming, costly and uncertain for contractors to get visas for workers, Pacific countries governments could ensure a streamlined, transparent, and fair process for the issuance of visas for infrastructure projects, working with MDBs and other donors to ensure efficient delivery of infrastructure projects.

(v)   Consider increasing bidding durations for tenders (currently a minimum of 4–6 weeks) to allow all stakeholders and potential bidders to fully understand supply chain risks and costs (materials, transportation, labor, insurance etc.) specific to the bid.

(vi)   Consider a Pacific country-wide adoption of standdown rates, meaning if the weather is not suitable for discharging cargo needed to deliver infrastructure projects, there would be an associated payment and/or time extension to a contractor.

These actions have the potential to assist governments, MDBs, and other bilateral donors in their efforts to attract and retain contractors to their future project pipelines. The actions can also achieve more long-term and cultural change that has the potential to substantially improve the situation.

Long-term Actions

(i)   To the extent possible, design infrastructure to reflect the environments in which they will be built and the logistical complexities inherent in their delivery (around both bulk and containerized freight). Engineering designs should be conscious of the need to maximize the use of local materials and skills while optimizing the transportation of imported plant, labor, and equipment to the islands without compromising on quality. This requires the engagement of consulting firms and contractors with a detailed knowledge of the region.

(ii)   Support the implementation of best practice. The perceived lack of a level playing field in construction supply chains in Pacific countries is a key issue. While the MDBs are not the clients, they can continue to advocate with Pacific countries to undertake procurement in accordance with best international standards. MDBs should continue to support Pacific countries in the move away from lowest-cost bidder approaches to tender evaluation towards a value for money approach that focuses more on delivering quality infrastructure aligned with Pacific country development outcomes. Improved communication between MDBs, Pacific country governments, and the construction market is required on an ongoing basis to remove the still-existing perception that all contracts are awarded on a lowest price basis.

(iii)   Increased regional standardization. MDBs and other regional donors should advocate strongly for, and provide direct support to, increased harmonization across the region including, but not limited to, the areas of occupational health and safety regulations, taxation and related exemptions for development projects, work visas, and digitization of financial systems to improve the efficiency and reliability of payment.

(iv)   Governments and development partners should coordinate and collaborate closely when planning their future project portfolios over the medium to long-term time horizon. This could lead help establish a sustainable and predictable pipeline of projects in Pacific countries that encourages increased international market interest, brings economies of scale in delivery, and reduces the potential for overloading local contractors and suppliers.

Guidance for Potential Contractors

Based on the results of our research that many of the risks become easier to manage with experience in this region, we recommend that new and potentially interested contractors strategically get "a foot in the door" with a couple of break-even projects and/or in joint venture with existing incumbent contractor delivering infrastructure across the region. While they may not initially make a sizeable profit from this exercise, it will enable them to acquire expertise and knowledge of the local market and donor funded project supply chain management context. In future rounds they can then bid to make profits and reap the potentially high financial rewards, in addition to the social benefits that come with these projects.

Further Research

Further investigation into the following areas is recommended.

(i)   Sharing risk vertically or horizontally. There appears to be interest in sharing risk. It would be valuable to explore whether this risk sharing should occur vertically or horizontally, that is, between local governments, MDBs, and suppliers (including transport) or at a level of any of these elements of the supply chain governance structure. For example, the idea of sharing a Pacific country freighter aircraft, shipping vessel, and/or barge pools has not been fully investigated. Who would own these assets and who would have access to the asset pool and under which contractual conditions (collaboration as a service) raises efficiency and equity issues. Such a risk pool could also be extended to insurance and other items that could be shared with the wider private sector.

(ii)   Exploring risk preferences and tolerances. It would be interesting to examine these through applied choice experiments with current and potential contractors. This would reveal preferences to accept risk and inform procurement and contracting of the vital components of successful supply chain risk management and enhance the broader understanding of the attractiveness of MDB-funded construction projects in the region.

(iii)   Investigating procurement. Transport service procurement through Pacific country governments and related competition effects (among transport companies but also contractors) are worth further investigation. Obtaining evidence on the effectiveness and efficiency of the practice deployed by the Micronesian shipping commission would be an excellent starting point.

(iv)   Assessing options for the importation of key materials like cement, steel, and aggregates across Pacific islands to reduce supply side risks, tariff and taxation uncertainty, and reduce project-related delays.

(v)   Exploring short-term and long-term risks of climate change. It would be interesting to explore the short versus long-term risks of climate change and how they interact with short-term interest related to large pipelines of MDB funded climate adaptation projects.

(vi)   Conducting more research on contracting options in a Pacific context. Allocating more work to a single contractor on a specific island would provide economies of scale, allow supply chains to mature, and increase planning certainty. However, it also raises market entry barriers and could lead to increased costs.

(vii)   Generating a matrix for Pacific countries where risk impact scenarios are to be adopted (e.g., valid for Tuvalu outer islands but not for Fiji etc.) and test how this can be used for pricing supply chain and contract management risks.

Appendix 1: Shipping Routes Across Pacific Island Countries in 2022

	Route 1	Route 2	Route 3	Route 4
Cook Islands (CI)	NZ – FJ- WS – TO- NU			
Federated States of Micronesia (FM)	PW – MH	PG – SB – FJ – MH		
Fiji (FJ)	NZ	PG – SB – MH – FM	NZ – FJ – TO - CI - NU	SB – VU – NC – WS – PF – TO - KI
French Polynesia (PF)	(AU)-NZ	NZ – TO	SB – VU – NC – FJ – WS – TO – KI	
Kiribati (KI)	FJ-TV	SB – VU – NC – FJ – WS – TO -PF		
Nauru (NR)	FJ			
New Caledonia (NC)	AU-VU-FJ	NZ – FJ – VU	NZ – PG – SB - AU	SB – VU – FJ – WS – TO – PF - KI
Niue (NU)	NZ – FJ – WS – TO - CI			
Palau (PW)	FM – MH			
Papa New Guinea (PG)	SB	AU - SB	NZ – NC – SB - AU	
Republic of Marshall Islands (MH)	PG – SB – FJ – FM			
Samoa (WS)	NZ – (AS)	PF – (AS)	NZ – FJ – TO - CI - NU	SB – VU – NC – FJ – TO – PF - KI
Solomon Islands (SB)	Pg	PG – FM – FJ – MH	NZ – NC – PG – AU	VU – NC – FJ – WS – TO – PF - KI
Tonga (TO)	NZ – PF	NZ – FJ – WS – CI – NU	SB – VU – NC – FJ – WS – PF – KI	
Tuvalu (TV)	FJ – KI	FJ – WS		
Vanuatu (VU)	NZ – NC - FJ	AU – NC - FJ	SB – NC – FJ – WS – TO – PF - KI	

Source: Jephcott (2022), p. 14.

Appendix 2: Merchandise Exports of Pacific Island Countries in 2022

Reporting Economy	2019 $ million	2020 $ million	2021 $ million	2022 $ million	2022 % of global	2022 Global Rank
Cook Islands	17	20	15	11	0.0000	201
Fiji	1,033	826	815	1,100	0.0044	151
Kiribati	12	9	9	12	0.0000	200
Marshall Islands	55	44	46	43	0.0002	189
Micronesia	49	50	76	134	0.0005	179
Nauru	19	72	121	117	0.0005	182
Niue	2	1	1	1	0.0000	209
Palau	4	7	2	3	0.0000	207
PNG	11,399	9,288	10,885	15,200	0.0610	86
Samoa	49	37	29	39	0.0002	190
Solomon Islands	461	379	371	333	0.0013	171
Tonga	20	15	16	16	0.0001	197
Tuvalu				1	0.0000	212
Vanuatu	56	46	54	64	0.0003	183
Pacific 14 total	13,176	10,794	12,440	17,073	0.0686	
World total	19,014,182	17,648,303	22,343,840	24,904,489		

Source: World Trade Organization (2023).

Appendix 3: Merchandise Imports of Pacific Island Countries in 2022

Reporting Economy	2019 m USD	2020 m USD	2021 m USD	2022 m USD	2022 % of global	2022 Global Rank
Cook Islands	136	105	119	124	0.0005	201
Fiji	2,734	1,731	2,116	2,984	0.0116	151
Kiribati	132	133	176	106	0.0004	204
Marshall Islands	68	74	81	94	0.0004	205
Micronesia	194	207	197	223	0.0009	196
Nauru	36	46	43	35	0.0001	207
Niue	14	12	14	16	0.0001	210
Palau	193	149	156	211	0.0008	197
PNG	3,934	3,289	3,383	3,189	0.0124	147
Samoa	391	312	368	435	0.0017	190
Solomon Islands	590	479	556	644	0.0025	183
Tonga	238	229	246	197	0.0008	198
Tuvalu	33	34	34	34	0.0001	208
Vanuatu	357	301	339	413	0.0016	191
Pacific 14 total	9,050	7,101	7,828	8,705	0.0340	
World total	19,338,838	17,881,471	22,620,211	25,621,162		

Source: World Trade Organization (2023).

Appendix 4: Interview Guide for Semi-Structured Interviews with Contractors

Amended versions were used for interviews with other stakeholders

Determinants of construction sector supply chain risks in Pacific countries

1. Kick-off Question: What are the key supply chain risks that you see for the Pacific Island construction sector more generally? (What supply chain risk [hazard]? Evaluate different types of supply chain (SC) risks in Pacific islands and how they affect your business.

    What about Staff? Workplace health and safety issues? Currency? Inflation? Delivery times/reliability? Scarcity? Scale? Geopolitics? Is the risk associated with

    (a)   certain islands, -routes, -modes of transport, -specific suppliers,
    (b)   COVID-19 (if so to what extend has the situation improved lately?),
    (c)   global SC bottlenecks (if so to what extend has the situation improved lately?),
    (d)   Climate change?

2. How much of your business construction portfolio is in the Pacific Islands (approx. in % of revenues)? Predisposition of your firm's SC to the risks identified in 2 (vulnerability = predisposition + sensitivity + probability + impact)?

3. Do you consider your business to be a tier 1, 2 or 3 supplier/contractor?

4. How experienced would you say is your firm with undertaking construction projects in Pacific islands? Does experience matter to managing SC risks in this region?

5. In which way is your firm exposed to SC risks in the Pacific islands? (infra, assets, operations, people, governance, brand, interdependencies)?

6. Are there existing protection measures to mitigate your firm's SC risk identified so far (level of preparedness and institutional capacity/adaptability)?

7.   Are there additional/different construction SC risks in donor-funded projects in Pacific islands?

  If so, how do they and your firm's response differ to any of the risks discussed in 1.

  (a)   What premium would a donor agency have to pay to make this risk acceptable/tolerable to you?
  (b)   What risk considerations does your firm evaluate when bidding for such projects? How much of those risks are SC related and how much do they contribute to a go/no-go decision?
  (c)   What other measures could a government body or project donor organization undertake to mitigate these risks (make them more tolerable)?
  (d)   How does your firm manage your SCs and logistics related to donor-funded construction projects in Pacific islands?
  (e)   What in your perspective are detriments for firms to engage more actively in construction projects in Pacific islands and to what extend do SC risks play a role in this? Are donor-funded projects different in that regard and if so in what sense?

8.   Are there any challenges or opportunities related to vertically integrated supply chains in this context?

9.   In terms of an outlook, over the next 6-12 months do you anticipate the construction SCs in Pacific island countries to:

  (a)   run far smoother than now
  (b)   run somewhat smoother than now
  (c)   run as smoot as now
  (d)   run somewhat less smooth than now
  (e)   run far less smooth than now?

10.   In summary, what are in your view the top current 5 supply chain risks for construction projects in Pacific islands?

11.   Do you anticipate each of those 5 SC risks to:

  (a)   Improve,
  (b)   remain the same,
  (c)   worsen over the next 12 months?

12.   In terms of looking further into the future, what do you think will be the top 5 SC risks for construction projects in Pacific islands in 10 years from now?

Appendix 5: Sample of Interviewed Stakeholders

	Organizations	Individuals
Contractors	8	11
Hall Contracting Pty Ltd.		
Reeves International		
Icon Construction		
Fletcher Construction Co. Ltd		
Downer New Zealand Ltd		
Pacific Marine Group Pty Ltd		
Kitano Construction Group		
McConnell Dowell Constructors Ltd		
Suppliers	4	4
Pacific Islands International		
Sino Soar Hybrid (Beijing) Technology		
Clay Engineering Pte Ltd		
Infratec NZ Limited		
Potential contractors	5	5
Bilfinger and Berger		
Lendlease		
Hochtief		
VINCI Construction		
Eiffage Société Construction General		
Potential suppliers	4	4
Knauf Gypsum		
Nippon Concrete Industries		
Daikin Industries		
Heidelberg Materials		
PICs shipping lines/ports	4	4
Swire Shipping		
Neptune Pacific Direct Line		
PNG Ports Corporation		
Fiji Ports Corporation		

continued on next page

Appendix 5 Table 5 *continued*

	Organizations	Individuals
PICs airlines/airports	4	4
Air Niugini		
Fiji Airways		
Fiji airport		
Nauru Airlines		
PICs Government authorities	5	5
Tonga Maritime and Port, Ministry of Infrastructure		
Solomon Islands Ministry of Infrastructure Development		
RMI Ministry of Transportation, Comm., and Information Technology		
Fiji Land Transport Authority		
Tuvalu Climate Change and Policy Unit		
PICs consultants/academics	3	3
International consultants/Engineers	3	3
DFAT, Austrade, MFAT, The Pacific Community (SPC)	4	8
Australia Pacific Island Business Council (+PNG + Fifi) council)	1	1
Multilateral donor banks (ADB/WB/IVB/PRIF/IMF/AIFFP/GCF)	7	7
Total	**52**	**59**

ADB = Asian Development Bank, AIFFP = Australian Infrastructure Financing Facility for the Pacific, GCF = Green Climate Fund, IMF = International Monetary Fund, MFAT = New Zealand Ministry of Foreign Affairs and Trade, NZ = New Zealand, PNG = Papua New Guinea, PIC = Pacific Island Countries, PRIF = Pacific Region Infrastructure Facility, WB = World Bank.
Source: ADB.

Appendix 6: Supply Chain Risks Specific to Multilateral Development Bank-Funded Construction Projects in Pacific Island Countries

Stakeholders interviewed for this study clearly recognize the benefits of projects funded by multilateral development banks (MDBs). These include transparent and fair procurement, availability of financing, increased certainty of payment, and the opportunity to contribute to the social and economic development of the Pacific countries, including climate change adaptation and transformation. While supply chains continue to normalize following the coronavirus disease (COVID-19) pandemic, MDB-funded projects come with their own challenges, many of which are avoidable if processes, governance structures, and perceptions and knowledge of contractors and Pacific country governments are improved.

There is appreciation that MDBs are not the clients, they only provide and broker the loans. The Pacific country governments are the clients. This results in various issues, starting with limited Pacific country government capacity to plan, and bureaucratic procedures, which combined can result delay awarding of contracts up to 18 months (time between the date of bid issuance and contract award). Contractors say governments also often have unrealistic time expectations them, even though the contractors are local, as they do not appreciate how much time and risk good procurement can take. Pacific country governments took a different view but admitted that when procuring MDB projects and supplies they are told to follow the Asian Development Bank (ADB) and World Bank procurement guidelines to accept the lowest complying price. Importantly, this issue is a problem of lagging perception, as this has not been a requirement since 2017 for ADB (2016 for World Bank) and it is stated earlier in the main body of this report that the majority of large value ADB tenders over the last 3 years have used quality-based evaluation methods to some degree. However, if Pacific countries have that perception and then award contracts to cost-leading bidders, contractor performance can sometimes be an issue. This is partly due to relying on a few contractors. For example, in Tonga, only three local engineering firms operate, and they work together in competition with some international bidders. Importantly, the perception is that the MDB bidding guidelines do not allow more expensive bidders (as noted above this is no longer the case in practice, so the issue is lagged perception of both Pacific countries governments and contractors). In addition, there is no feedback or information on whether bidders have delayed previous projects or had quality and performance issues.

Issues that May Lead to Lack of Bidders

Another key finding is that while MDBs are interested in attracting more contractors and bidders for their very large pipelines of projects and other work coming out over the next few years,[1] current contractors said more attention should be given to them. They have many recommendations for improving the process, commercial terms, and governance structures around MDB-funded construction projects in Pacific countries, including aiming for a more level playing field. They fear that without those changes, even fewer bidders will come to the market. They would welcome a face-to-face roundtable with all stakeholders, and noted that doing this

---

[1]   On top of the projects funded through the A$4 billion ($2.6 billion) Australia Infrastructure Financing Facility for the Pacific and other MDB pipelines and local government-funded work.

through business councils would perhaps be more neutral. Five leading contractors that work in the region and interviewed for this report would provide firsthand advice on key issues.

The "Developing Sustainable Resilient Infrastructure in the Blue Pacific conference" took placed on 25–27 September 2023 in Brisbane, Australia. Hosting the event were the Australia Pacific Islands Business Council in collaboration with the Australia Fiji Business Council and the Australia Papua New Guinea Business Council, supported by the Government of Australia's Australian Infrastructure Financing Facility for the Pacific. This provided a good opportunity to connect with current and potential contractors. However, interviewees suggested that every stakeholder also needs smaller meetings to help build trusted relationships. This could help ensure that donors and governments do not wait for the lowest bid and price to come into the tender box, but instead engage early with (potential) contractors to realize projects that make a difference to the region and that are commercially of interest to a wide range of contractors and suppliers.

From a Pacific country government perspective, especially from medium and smaller islands, it was further mentioned that donors require them to employ international experts and over rely on international engineers for large projects. Reportedly (mentioned by three interviewees), the lead engineer is normally international and will only be there for a short time or on an intermittent basis, as the Pacific country governments cannot afford more time. One idea raised was to train more locals and to request available local engineers. MDBs commented that they do advocate for local expertise where available and that incorporating knowledge transfer requirements into contracts with international consultants is increasingly being explored.

From the MDB perspective, there is uncertainty as to the optimal size of construction projects, especially for transport infrastructure. For example, many World Bank contracts in transport are $1 million–$5 million at the smaller end of the scale and the supply chain risks are too large to attract reasonably sized and experienced companies, as they will chase bigger projects. As such, the donor agency is often stuck in the middle, and may want to try to relax financial and cash flow requirements to encourage more local contractors to bid. While ADB, the World Bank, the Australian Department of Foreign Affairs and Trade (DFAT), New Zealand's Ministry of Foreign Affairs and Trade (MFAT), and Austrade undertake business seminars that present their pipeline of upcoming projects, they find it difficult to attract new market entrants as these events are predominantly attended by consultants and contractors already working in the region.

Australian and New Zealand government agencies that aim to support the economic development of and trade with Pacific countries, acknowledge that they only have a small pool of contractors operating in the region with whom they try to build trusting partnerships. Reportedly they are trying to be "extra kind" to contractors from a risk sharing and award perspective, as they see value in keeping them in the market to have at least some level of competition. As the government agencies recognize some of the supply chain and logistics issues listed in Table 3, they are now seeking alternative modes of engaging, such as early contractor involvement where they engage preferred contractors in the early design stage of building and where logistics issues are discussed early with local building experts.

To secure sustainability and economic development, DFAT and MFAT now place more value on procurement and have started to add circular supply chains and zero waste and heavy involvement of local labor (self-contained) to contract requirements, with local people to be trained up, and food and direct supplies bought locally. Most interviewed contractors indicated they support the desire to build up local teams and manage local assets. Bids are advertised internationally on a website and there is some new interest with two new Australian tier 2 companies bidding last year) but no new international (non-Australia, New Zealand or Asian newcomers). In addition, tender documentation and evaluation criteria have high weights on experience in working in the Pacific (also part of the two-stage bidding process of the MDBs), as the unique operating environments are acknowledged due to local conditions, remoteness and geography not only being different from Australia or New Zealand but also within the Pacific countries.

Once projects have commenced, sometimes DFAT/MFAT agencies have to step in and contract Australian or New Zealand firms and shipping agents (e.g., Alaska based shippers) for transport tasks to further support an already-mobilized contractor. Some contractors initially take on these risks but push back later claiming force majeure grounds, as they argue that they cannot manage many of the supply chain risks in the Pacific Island context, such as local island ferries being out of service for months. DFAT/MFAT then often pays the cost for alternative logistics methods or pays variation charges. That said, both agencies commented that contractors are innovative in finding solutions and working with them when supply chains are disrupted and that they are usually well-connected. It is still problematic as they cannot predict how much the disruption will cost both in terms of money and time. It was further noted that in case of disruptions, there are not a lot of vessels that can access some Pacific countries, which results in overreliance on Australian or New Zealand assets.

From a contractor perspective, experienced and large construction companies prefer MDB-funded projects, as these contractors are big enough to meet the qualification requirements and know that payment will be forthcoming. Interviewees mentioned that sometimes local Pacific country governments financed projects do not pay, as money is not available, or they pay very late which can result in some contractors only bidding for MDB-funded projects.

Cash Flow and Payment Issues

While a principal reason for why contractors bid on MDB projects is cash availability, another supply chain risk frequently mentioned during the interviews was cash flow issues with donor organizations. Procurement and payment terms can be archaic with 90 days being the norm, but it can stretch out to 120 days. Some contractors said that donor payment systems are not up to standard. An issue arises because the donor agencies are not the client and there is often no direct line between the contractor and donor organization, which leaves the contractor having to deal with the government of a developing country. In some instances, where governments systems are assessed as being weak, MDBs may disburse funds to contractors and suppliers, but this only occurs after the government has requested the MDB to make a payment. In one example provided, on numerous occasions the government in Papua New Guinea had to sign off on payments which were then delayed due to lost information and capability; in other cases, Pacific countries changed items on the invoice which the World Bank or ADB then could not match with their system and the invoice then does not get paid or is severely delayed. The contractor has to pay everything in advance and thereby takes on a high risk for payment delays which need to be financed, with 90 day-bank guarantees or funds in retention for up to 18 months. The highly experienced firms say that it will be almost impossible to attract other contractors given this issue, let alone all the other supply chain risks.

The payment delay risk then also rolls into supply chains as contractors need to pay suppliers well before then. If quality issues are picked up after the supplier has been paid, it is too late, and often components do not arrive for 3 months and then it is another 3 months to wait for payment. This was not problematic when interest rates were around 1%, but with rates now at 6%–8% it is costly and presents a major risk that contractors must consider before bidding for contracts. There is also often a lack of digitization of payments once a performance milestone has been met. This is reportedly much worse in projects directly procured by Pacific governments as they lack the technology and capacity to do this. In many cases, payment is still based on paper systems.

Contractors often need to pay a temporary import bond to get materials, plant, and equipment into Pacific countries. Some argue that exemptions, which seem to be granted to some but not all, would help. Contractors also need to pay wages. For example, in Papua New Guinea, there is a much higher income tax for foreigners and contractors which then results in contractors needing to compensate staff and top up salary: 45% flat rate income tax, 15% foreign contractor withholding tax, combined with a 10% goods and services tax at home in Australia, and import duty to protect Pacific Island businesses, etc. One suggestion was that aid infrastructure should be tax exempt.

Premiums and Mechanisms to Address Risks

When asked what premium a donor agency would have to pay to make the supply chain risks acceptable and tolerable to contractors, there were a wide range of responses.

Three interviewed contractors stated that they would normally charge a premium or build risk into the contract depending on the Pacific country or the project specific risk. However, they acknowledge that this impacts their competitiveness, particularly if price is the defining award criterion. A fourth contractor advised that the key objective should be to share risks, which can work out well but can also work out poorly for both sides. It appears that most contractors believe that without the use of collaborative approaches, the risks outweigh the benefits, particularly given the amount of work they can get in other jurisdictions. Our interviews suggest that all stakeholders agree that early contractor involvement should be considered on projects, particularly in remote and high-risk locations.

A further risk mitigation strategy to reduce the exposure to cash-flow risks on donor-funded construction projects was increased use of advance payments for plant and materials. The following is a brief example to show how this works. Say the contractor submits a quote and places an order for 200 tons of bitumen worth A$4 million ($2.6 million) but, at the same time, they say to the client they cannot carry the cash flow risk. The client agrees to pay the actual cost when the order is placed as an advance payment and a bond. The risk, therefore, moves to the client or donor and is related to extreme foreign exchange volatility, commodity (i.e., fuel and steel) and shipping price volatility. However, this works both ways, as the client may profit if commodity prices fall until delivery of the materials. The benefit of this approach is that it removes the risk exposure from the contractor and makes the risk fully transparent. However, reportedly MDBs are not fully aligned with the idea as they do not like fluctuations either. The alternative is limited competition across the region which will likely lead to increased costs to deliver projects and potential reduction in achieving the delivery of quality infrastructure.

A further identified issue is that the donor budgets for project development can be insufficient which, in the view of contractors, results in engineering consultants who are inexperienced undertaking the work with limited understanding of the complexities and unique operating environment in the Pacific. Limited funding in the development phase can also lead to insufficient analysis of ground conditions leading to a preference to pass these risks onto contractors at construction stage. An example of this reported by interviewees was on the A$4 billion ($2.6 billion) Australia Infrastructure Financing Facility for the Pacific facility where two tenders failed due to unbalanced risk allocation leading to added risk premiums and greatly increased costs. Earlier contractor involvement is seen as a good way to overcome some of these challenges allowing experienced contractors to be involved early in the process while at the same time creating an environment of increased trust between the parties. MFAT has successfully tested this approach in Tokelau, where the lack of transport infrastructure was a severe constraint, and the contracting industry was engaged at an early stage to assist in the design of the project.

Potential and current contractors also mentioned that project bidding costs are high. For example, for a contractor to price a $50 million project, the tender cost would be around $50,000-$100,000 plus a substantial time investment. The perception of a limited possibility of success, aligned with concerns regarding competitors who may undercut the opposition to win contracts can lead to firms not engaging in the process. In addition to informing bidders of the MDB's moves to increasing use of quality-based evaluation methods, a suggested solution for this risk is to reimburse some or all of the costs incurred in developing high-quality tender submissions. To make this work, there needs to be robust oversight on how the final decision for project award is made and a transparent system for validating high quality bids detailed in the bidding documents.

Finally, it was suggested that contractors should be able to use aggregate construction materials sourced locally in the Pacific countries and that bidding agencies should be more flexible in contract procurement to achieve fit for purpose performance specifications. To increase market interest and thus competition, contractors also proposed the establishment of a roster of highly qualified and experienced contractors for the delivery of the substantial portfolio of Pacific projects. This would increase visibility on competition and provide confidence to companies to bid on infrastructure projects.

For potential contractors, these issues are compounded by the prohibitive cost to mobilize and set up businesses on the Pacific islands.

Findings

Our analysis of supply chain risks specific to MDB-funded construction projects in Pacific countries suggests that most potential bidders from regional countries are not very interested in accepting many of the risks identified in this report. There is so much work in their home jurisdictions and markets and a skill shortage in engineers and talented staff globally that they prefer to concentrate on markets they are familiar with. A key additional problem is that there is lagged perception of Pacific country governments and contractors around MDB procedures, guidelines, and contractual mechanisms for risk mitigation.

Potential contractors from entirely new markets (Europe, United States, and Japan) suggested that they would prefer to expand to safer and institutionally more developed countries, such as Australia and New Zealand first, where there is a lot of work that carries less risk. Only once they have established an operational base in those countries would they consider branching further out into the Pacific. There are some activities by European and Japanese authorities to assist their companies to engage more in the development of the Pacific, but no solid interest of potential contractors from either of these jurisdictions was obtained through the interview program or the Austrade Pacific Island International Development Opportunities event. As such, in the current institutional and economic environment, it is going to be challenging to attract new contractors to Pacific Island construction projects.

Perhaps more importantly, our research identified several unhappy current contractors and suppliers, who may soon be forced out of this market and/or be unable and/or unwilling to compete for further contracts. MDBs and other donors should focus on retaining current contractors to avoid an already challenging procurement environment from becoming worse. Many contractors are keen on one-on-one dialogue with MDBs to find solutions that resolve the current challenges (e.g., transport delays, customs clearance bottlenecks, increased and volatile costs, and logistical complexities) they see in delivering projects in the Pacific. Given the importance of a smooth and efficient supply chain to their overall operation, they believe it is imperative to address these matters proactively and collaboratively.

Recommendations

Short term

1. Contractors would welcome a face-to-face roundtable with all stakeholders at the table to discuss project pipelines, implementation challenges and associated risks. This should be facilitated and then maintained on a regular basis.

2. MDBs should advance the use of collaborative forms of contracting such as early contractor involvement and competitive dialogue to incorporate the knowledge and experience of all stakeholders in the design of Pacific infrastructure to increase the possibility of project success.

3. Working with partners like MDBs and other donors, Pacific country governments should improve payment systems. It should be relatively easy to improve the MDB-funded payment systems and platforms to enhance the speed and reduce the bureaucracy and uncertainty around milestone payments. MDBs could support this work through additional technical assistance and provision of information on best international practices. This process should also reconsider payment terms and how these affect contractor and supplier risks and financing costs.

4. For very specialized projects, MDB-funded safeguard and procurement systems and contracts could allow for more flexibility to better reflect the local context and need for specialist know-how and equipment.

5. Given the lack of awareness among Pacific countries and contractors, communicate the existence of risk and cost sharing tools available within the procurement policies of donors and MDBs. Explore more active risk mitigation and cost sharing tools (internal systems but also with external partners such as insurance brokers or investment banks) such as fuel, interest rates and foreign exchange hedging that could be shared with either the contractors or the clients (Pacific country governments) or both. Increased use of advance payments and/or guaranteeing or fixing material prices in the local currency (to reduce the US dollar risk) should be part of such considerations.

6. Pay for bids. Paying bidders for large projects a fixed compensation fee for solid bids may increase the number of overseas bidders. In Australia, compensating bid costs is common for construction projects of over A$100 million ($65 million). This recommendation could be the subject of additional research on best international practices.

7. Increase time and payment for engineers. MDBs should consider slightly increasing the payments for bid engineering consultants and more importantly increase the time they have available to do what they are asked in a more diligent and appropriate way. Engineers with experience in the region should be preferred.

Long term

1. Increase funding allocations across the project life cycle to include operations and maintenance activities. As funds from donor agencies are large lump sums, this creates supply chain and cash flow risks and does not incentivize stakeholders to complete projects to a high quality. This reportedly results in many cases of assets that do not fulfil the intended purpose. One option to address this issue would be to engage contractors for both the capital construction and maintenance of assets. MDBs could finance projects where the Pacific country government enters into a single contract for design, construction, maintenance, and operation of an infrastructure facility over a contractually defined period. Structured appropriately, the contractor would have a vested interest in constructing a facility that requires minimum maintenance during its operational life.

References

Alexander, D. and R. Merkert. 2017. Challenges to domestic air freight in Australia: Evaluating air traffic markets with gravity modelling. *Journal of Air Transport Management*. 61. pp. 41–52.

Al-Mhdawi, M.K.S., M. Brito, B.S. Onggo, A. Qazi, A. O'Connor, and B.M. Ayyub. 2023. A Structural Equation Model to Analyze the Effects of COVID-19 Pandemic Risks on Project Success: Contractors' Perspectives. *ASCE-ASME Journal of Risk and Uncertainty in Engineering Systems, Part A: Civil Engineering*. 9 (31). 05023003.

Choudhary, N.A., S. Singh, T. Schoenherr, and M. Ramkumar. 2023. Risk Assessment in Supply Chains: A State-Of-The-Art Review of Methodologies and their Applications. *Annals of Operations Research*. 322. pp. 565–607.

Dyer, J. 2017. Adapting Climate Change Projections to Pacific Maritime Supply Chains. In W. Leal Filho, ed. Climate Change Adaptation in Pacific Countries, *Climate Change Management*. Springer, Cham.

Gurbuz, M.C., O. Yurt, O. Ozdemir, V. Sena, and W. Yu. 2023. Global Supply Chains Risks and COVID-19: Supply Chain Structure as a Mitigating Strategy for Small and Medium-Sized Enterprises. *Journal of Business Research*. 155, Part B. 113407.

International Bank for Reconstruction and Development/World Bank. 2023. *Connecting to Compete 2023 – Trade Logistics in an Uncertain Global Economy – The Logistics Performance Index and Its Indicators*.

Jephcott, G. 2022. Investigation of the Supply-Chain Disruption Due to the Pandemic and Its Economic Impacts on Business across the Forum Island Countries, including Micro, Small and Medium Enterprises. Study Commissioned by Pacific Trade Invest Australia, Pacific Trade Invest New Zealand, and Pacific Island Forum.

Merkert, R., K. Hoberg, and K. Mahadevan. 2023. What Are the Logistics and Supply Chain Superpowers and Skills to Survive in the "New Normal" Globalized World? *Transportation Journal, Special Issue: Teaching 'Em Up: Preparing Tomorrow's Supply Chain Leaders* (forthcoming).

Niroshana, N., C. Siriwardana, and R. Jayasekara. 2022. The Impact of COVID-19 on the Construction Industry and Lessons Learned: A Case of Sri Lanka. *International Journal of Construction Management*. DOI:10.1080/15623 599.2022.2076016.

Pacific Region Infrastructure Facility (PRIF). 2024. Insurance Risk Management and Insurance in the Pacific. Consultant's report. PRIF, Sydney.

Panova, Y. and P. Hilletofth. 2018. Managing Supply Chain Risks and Delays in Construction Projects. *Industrial Management & Data Systems*. 118 (7). pp. 1413–1431.

Pham, H.T., T. Pham, H.T. Quang, and C.N. Dang. 2023. Supply Chain Risk Management Research in Construction: A Systematic Review. *International Journal of Construction Management*. 23 (11). pp. 1945–1955.

Rao, S. and T.J. Goldsby. 2009. Supply Chain Risks: A Review and Typology. *The International Journal of Logistics Management*. 20 (1). pp. 97–123.

Rinaldi, M. and E. Bottani. 2023. How Did COVID-19 Affect Logistics and Supply Chain Processes? Immediate, Short, and Medium-Term Evidence from Some Industrial Fields of Italy. *International Journal of Production Economics*. 262. 108915.

Rokooei, S., A. Alvanchi, and M. Rahimi. 2022. Perception of COVID-19 Impacts on the Construction Industry Over Time. *Cogent Engineering*. 9 (1). DOI:10.1080/23311916.2022.2044575.

Thompson, J. 2022. Mitigating Risk in the Construction Supply Chain. *Infrastructure Magazine*. 30 November. https://infrastructuremagazine.com.au/2022/11/30/mitigating-risk-in-the-construction-supply-chain/

UNESCAP. 2022. Status Report on Sustainable and Resilient Ports and Maritime Connectivity in the Pacific Region. https://www.unescap.org/sites/default/d8files/event-documents/Sustainable_and_resilient_port_in_the_Pacific_April2022.pdf.

World Bank. 2023a. *Supply Chain Management, An introduction and Practical Toolset for Procurement Practitioners.*

World Bank. 2023b. *World Bank Procurement Approaches: Overcoming Supply Chain Challenges in the Pacific.*

World Trade Organization. 2023. WTO Stats Dashboard, World Trade Organization. https://stats.wto.org/dashboard/merchandise_en.html.

Zighan, S. 2022. Managing the Great Bullwhip Effects Caused by COVID-19. *Journal of Global Operations and Strategic Sourcing*. pp. 28–47.